MW01010020

TURBOMACHINERY

MECHANICAL ENGINEERING

A Series of Textbooks and Reference Books

Editor

L. L. Faulkner

*Columbus Division, Battelle Memorial Institute
and Department of Mechanical Engineering
The Ohio State University
Columbus, Ohio*

Additional Volumes in Preparation

Finite Elements: Their Design and Performance, Richard H. MacNeal

Computer-Aided Graphics and Design: Third Edition, Revised and Expanded, Daniel L. Ryan

Mechanical Engineering Software

Spring Design with an IBM PC, Al Dietrich

Mechanical Design Failure Analysis: With Failure Analysis System Software for the IBM PC, David G. Ullman

TURBOMACHINERY

BASIC THEORY AND APPLICATIONS

SECOND EDITION, REVISED AND EXPANDED

EARL LOGAN, JR.

Arizona State University
Tempe, Arizona

CRC Press
Taylor & Francis Group
Boca Raton   London   New York

CRC Press is an imprint of the
Taylor & Francis Group, an informa business

FIRST INDIAN REPRINT, 2014

Library of Congress Cataloging-in-Publication Data

Logan, Earl.
  Turbomachinery : basic theory and applications / Earl Logan, Jr. --
2nd ed., rev. and expanded.
      p.   cm. -- (Mechanical engineering ; 85)
  Includes bibliographical references and index.
  ISBN 0-8247-9138-X
  1. Turbomachines.   I. Title. II. Series: Mechanical engineering
(Marcel Dekker, Inc.) ; 85.
TJ267.L6 1993
621.406--dc20   93-1185
                                                                    CIP

The publisher offers discounts on this book when ordered in bulk quantities.
For more information, write to Special Sales/Professional Marketing at the
address below.

Copyright © 1993 by MARCEL DEKKER, INC. All Rights Reserved.

Reprinted 2009 by CRC Press

Printed and bound in India by Bhavish Graphics.

Neither this book nor any part may be reproduced or transmitted in any form
or by any means, electronic or mechanical, including photocopying, micro-
filming, and recording, or by any information storage and retrieval system,
without permission in writing from the publisher.

MARCEL DEKKER, INC.
270 Madison Avenue, New York, New York 10016
ISBN 978-0-8247-9138-4

FOR SALE IN SOUTH ASIA ONLY

Preface to the Second Edition

Use of the first edition during the past twelve years has led to the development of supplementary material for classroom instruction. This material has been integrated into the present edition.

The new material comprises equations, graphs, symbol lists, and illustrative examples that clarify the theory and demonstrate the use of basic relations in performance calculations and design. Additionally, a large number of problems has been added in Chapters 2 through 8. Most of these problems were developed as numerical or analytical exercises; however, a few were generated for design projects. In the latter case, the designs correspond to existing hardware for which performance data and dimensions are available.

Some material used in the first edition was relocated in the second edition and some was expanded to provide a complete, but concise picture of current knowledge useful in preliminary design. The axial-flow pump material from Chapter 6 has been relocated to Chapter 4. The axial-flow fan and compressor material has been combined in Chapter 6. Chapter 7 treats radial-flow gas turbines only and is an expansion of material formerly in the single gas-turbine chapter. Chapter 8 was enlarged and handles axial-flow gas turbines exclusively.

The reorganization has evolved from attempts to find packages of material suitable for classroom instruction that are optimal in both size and content. Although the book was spawned from classroom instruction of fourth-year engineering students, it can be used by practicing engineers outside the classroom and by engineering technology students in the classroom envirnoment. It is assumed that the student or practicing engineer has studied basic fluid mechanics and thermodynamics. With this academic background, one should be able to undertake and complete a study of this volume.

There is sufficient material in the first eight chapters for a one semester course. It is recommended that the first four weeks be devoted to Chapters 1 through 3, three weeks to Chapter 4, and two weeks each to Chapters 5, 6, 7, and 8. Experience with the material presented in this edition has indicated that, after 15 weeks of instruction in a three-semester-hour course, students are able to perform a very satisfactory preliminary design of any of several types of turbomachines.

Earl Logan, Jr.

Preface to the First Edition

This book is intended as a text for undergraduate students of engineering. The student should have had introductory courses in fluid mechanics and thermodynamics prior to taking a course for which this book is the assigned text. The book should also find usefulness in the reference libraries of graduate engineers who may not have studied the subject while in school.

The plan of the book is first to present the basic principles of turbomachine theory and then to apply these principles to specific devices. The centrifugal pump is studied first, since it is both a common and a simple device. Then more complex machines are considered, concluding with the basic types of hydraulic turbines. Each chapter seeks to address the questions of how the principles may be applied in design and how they may be used to predict the performance of the turbomachine under consideration. A conscious effort to minimize theoretical detail has been made, with the objective of maintaining a clear and unified exposition of the basics. References are cited wherein the student may find more information, if so desired. The contribution of the problems on wind turbines by Professor Robert H. Kirchhoff of the University of Massachusetts is gratefully acknowledged. The author is indebted to many teachers and students who have, over the years, shed light on the subjects discussed. Many sources have been drawn from in preparing the text, and these sources have been cited at appropriate places. Appreciation is expressed to the Arizona State University, and in particular, to Drs. George C. Beakley, Warren Rice, and Darryl E. Metzger, who during their tenure as Chairmen of Mechanical Engineering have permitted me to teach the subject matter contained herein.

Earl Logan, Jr.

Contents

TURBOMACHINERY

1 Types of Turbomachines

1.1 Introduction

Turbomachines constitute a large class of machines which are found virtually everywhere in the civilized world. This group includes such devices as pumps, turbines, and fans. Each of these has certain essential elements, the most important of which is the *rotor,* or rotating member. There is, of course, attached to this spinning component a substantial *shaft* through which power flows to or from the rotor, usually piercing a metallic envelope known as the *casing.* The casing is also pierced by fluid-carrying pipes which allow fluid to be admitted to and carried away from the enclosure bounded by the casing. Thus a turbomachine always involves an energy transfer between a flowing fluid and a rotor. If the transfer of energy is from rotor to fluid, the machine is a pump, fan, or compressor; if the flow of energy is from the fluid to the rotor, the machine is a turbine.

The purpose of the process described above is either to pressurize the fluid or to produce power. Useful work done by the fluid on the turbine rotor appears outside the casing as work done in turning; for example, it can turn the rotor of a generator. A pump, on the other hand, receives energy from an external electric motor and imparts this energy to the fluid in contact with the rotor, or impeller, of the pump.

The effect on the fluid of such devices is that its temperature and pressure are increased by a pumping-type turbomachine, and the same properties are reduced in passage through a work-producing turbomachine. A water pump might be used to raise the pressure of water, causing it to flow up into a reservoir through a pipe against the resistance of frictional and gravitational forces. On the other hand, the pressure at the bottom of a reservoir could be used to produce a flow through a hydraulic turbine, which would then produce a turning moment in the rotor against the resistance to turning offered by the connected electric generator.

1.2   Geometries

A typical turbomachine rotor, a centrifugal pump impeller, is shown schematically in Figure 1.1. Liquid enters the eye E of the impeller moving in an axial direction, and then turns to a radial direction to finally emerge at the discharge D having both a radial and a tangential component of velocity. The vanes V impart a curvilinear motion to the fluid particles, thus setting up a radial centrifugal force which is responsible for the outward flow of fluid against the resistance of wall friction and pressure forces.

The vanes of the rotor impart energy to the fluid by virtue of pressure forces on their surfaces, which are undergoing a displacement as rotation takes place. Energy from an electric motor is thus supplied at a constant rate through the shaft S which is assumed to be turning at a constant angular speed.

If the direction of fluid flow in Figure 1.1 is reversed, the rotor becomes part of a turbine, and power is delivered through the shaft S to an electric generator or other load. Typically, hydraulic turbines have such a configuration (see Figure 1.4) and are used to generate large amounts of electric power by admitting high-pressure water stored in dams to the periphery of such a rotor. A pressure drop occurs between the inlet and the outlet of the

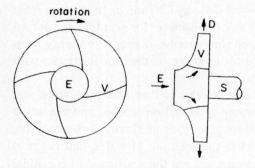

Figure 1.1 Pump impeller.

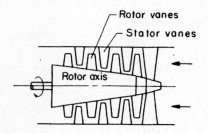

Figure 1.2 Axial-flow blower.

turbine; the water exits axially and is conducted away and discharged at atmospheric pressure.

If the substance flowing through the impeller of Figure 1.1 were a gas, then the device would be a centrifugal compressor, blower, or fan, depending on the magnitude of the pressure rise occurring during transit from inlet to outlet. For the reversed flow case, i.e., a radially inward flow, the machine would be called a radial-flow gas turbine or turboexpander.

A different type of turbomachine is shown in Figure 1.2. Here the flow direction is generally axial, i.e., parallel to the axis of rotation. The machine shown in this figure represents an axial-flow compressor or blower, or with a different blade shape an axial-flow gas or steam turbine, depending on the direction of energy flow and the kind of fluid present.

In all of the machines mentioned thus far, the working fluid undergoes a change in pressure in flowing from inlet to outlet, or vice versa. Generally, pressure change takes place in a diffuser or nozzle, and in the rotor as well. However, there is a class of turbines in which pressure change does not occur in the rotor. These are called impulse, or zero-reaction, turbines, as distinguished from the so-called reaction turbine, which allows a pressure decrease in both nozzle and rotor. A hydraulic turbine with zero reaction is shown in Figure 1.3, and a reaction-type hydraulic turbine appears in Figure 1.4.

Centrifugal machines are depicted in Figure 1.5 through 1.7, and axial-flow turbomachines are indicated in Figures 1.8 through 1.10. A mixed-flow pump is shown in Figure 1.11. This class of machine lies part way between the centrifugal, or radial-flow, types and the axial-flow types.

Sizes vary from a few inches to several feet in diameter. Fluid states vary widely as well. Steam at near-critical conditions may enter one turbine, while cool river water enters another. Room air may enter one compressor,

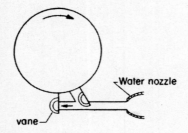

Figure 1.3 Pelton wheel.

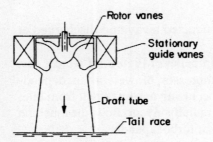

Figure 1.4 Francis turbine.

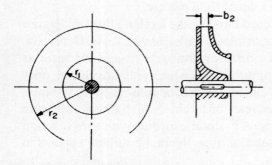

Figure 1.5 Centrifugal pump.

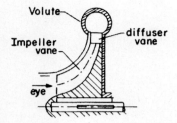

Figure 1.6 Centrifugal compressor.

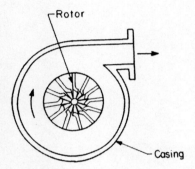

Figure 1.7 Centrifugal blower.

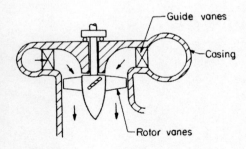

Figure 1.8 Kaplan turbine.

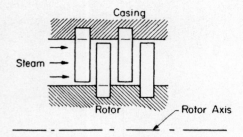

Figure 1.9 Steam turbine.

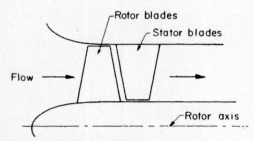

Figure 1.10 Axial-flow compressor.

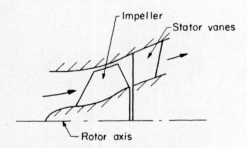

Figure 1.11 Mixed-flow pump.

while cold refrigerant is drawn into a second. The materials encountered in the machines are selected to suit the temperatures, pressures, and chemical natures of the fluids handled, and manufacturing methods include welding, casting, and machining.

Our consideration herein of the subject of turbomachines includes a wide variety of forms and shapes, made of a variety of materials using a number of techniques. This book does not attempt to deal with all the problems encountered by the designer or user of turbomachines, but only with the most general aspects of the total problem. The present treatment is concerned with specification of principal dimensions and forms of those turbomachines encountered frequently in industry.

1.3 Practical Uses

The importance of the turbomachine to our way of life cannot be over-emphasized. The steam power plant, which is responsible for the generation of most electrical power in the world, can be used to illustrate this basic fact. The steam power plant consists of a prime mover driving a large electric generator. A steam turbine is usually used as the prime mover. Steam for the turbine is supplied from a boiler at high pressure and temperature. Water for making the steam is forced into the boiler by means of a multiple-stage centrifugal pump. Fuel for creating the heat in the boiler is supplied by a pump, compressor, or blower, depending on the nature of the fuel. Air for combustion of the fuel enters the boiler through a large centrifugal fan. After the steam has been generated in the boiler and has expanded in the turbine, it is exhausted into a condenser where it is condensed and collected as condensate. Pumps are used to remove the condensate from the condenser and deliver it to feedwater heaters, from which it is drawn into the boiler feed pumps to repeat the cycle. The condensation process requires that large amounts of cooling water be forced through the tubes of the condenser by large centrifugal pumps. In many cases the cooling water is itself cooled in cooling towers, which are effective because large volumes of outside air are forced through the towers by axial-flow fans.

Thus we see that many turbomachines are required to operate the simplest form of modern steam-electric generating station. It is clear that modern industry and the entire economy depend upon such generating stations, and hence we are all dependent upon turbomachines in this and in many other applications.

Before we consider the specifics of pumps and turbines, we must deal with the fundamentals underlying their design and performance. In Chapters 2 and 3 we will develop these underlying principles of all turbomachines by starting our discussion with the first principles of physics, i.e., conservation of mass, momentum, and energy.

2   Basic Relations

2.1   Velocity Diagrams

The rotor shown in cross section in Figure 2.1 will have fluid flowing in the annulus bounded by abcda. Although fluid velocity varies radially from a to b, it is assumed to have a single value over the entire annular section ab, namely, the velocity V_1 at point 1. Similarly, at the rotor exit the velocity V_2 is taken as the average of the velocity along cd. Points 1 and 2 lie on the line 1–2 which denotes an element of a stream surface which exactly divides the flow into two equal parts.

Figure 2.2 shows a velocity diagram at point 1. The blade, or vane, velocity is calculated from

$$U = Nr \tag{2.1}$$

for any point on the blade a radial distance r from the axis A–A of rotation. The angular speed (rad/s) of the rotor is denoted by N. For point 1, (2.1) becomes

$$U_1 = Nr_1 \tag{2.2}$$

2.2   Mass Flow Rate

The relative velocity W of the fluid, with respect to the moving vane, is added vectorially to the blade velocity U to obtain the absolute fluid velocity V. The relation can be expressed by

9

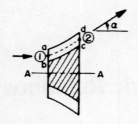

Figure 2.1 Turbomachine rotor.

$$V = W + U \tag{2.3}$$

The graphical representation of the addition of U_1 to W_1 is shown in Figure 2.2. The corresponding velocity diagram at the rotor outlet is shown in Figure 2.3.

The mass flow rate $\dot{m}$ through the rotor is calculated by multiplying the meridional velocity V_m by the area (normal) of the flow passage and by the fluid density. For example, at the rotor inlet

$$\dot{m} = \rho_1 A_1 V_{m1} \tag{2.4}$$

When the flow direction is at an angle to the rotor axis, a more complicated expression is obtained, but the same principle applies. The general form of (2.4), applying to any station in the flow passage, is

$$\dot{m} = \rho V_m A \tag{2.5}$$

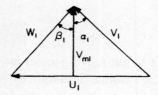

Figure 2.2 Velocity diagram at inlet.

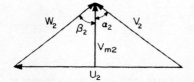

Figure 2.3 Velocity diagram at outlet.

where A is the area normal to the flow direction. Equation (2.5) is a statement of conservation of mass, i.e., the mass flow rate is the same at all stations.

It is assumed that each velocity on the central streamline of the flow passage depicted in Figure 2.1 is the average value for the entire flow area at the position considered. The actual flow has a variable velocity across the passage. The variation can, of course, be handled mathematically through the use of the integral form

$$\dot{m} = \int_A \rho V_m \, dA$$

(2.6)

The latter form also allows for possible variations in density associated with temperature, pressure, or concentration gradients.

2.3 Energy Equation

The principle of conservation of mass, expressed in (2.5), must be supplemented by a steady-flow energy equation which expresses the conservation of energy. The usual forms of energy per unit mass, which must be accounted for in a turbomachine, are potential energy zg, internal energy e, flow work p/ρ, kinetic energy $V^2/2$, heat transfer q, and work w. A word statement of the energy equation is the following:

Energy at section 1 + heat transfer = energy at section 2 + work done between 1 and 2

In equation form this is written as

$$z_1 g + E_1 + \frac{p_1}{\rho_1} + \frac{V_1^2}{2} + q = z_2 g + e_2 + \frac{p_2}{\rho_2} + \frac{V_2^2}{2} + w \tag{2.7}$$

Such an equation has been derived in many thermodynamics treatises, such as that by Jones and Hawkins (1986). Frequently, the internal energy is combined with the flow work to form the enthalpy h. The equation is then

$$z_1 g + h_1 + \frac{V_1^2}{2} + q = z_2 g + h_2 + \frac{V_2^2}{2} + w \tag{2.8}$$

Usually, in turbomachinery applications the potential energy and the heat transfer terms are neglected, and the specific work is denoted by E and called energy transfer, with the result

$$h_1 + \frac{V_1^2}{2} = h_2 + \frac{V_2^2}{2} + E \tag{2.9}$$

In gas turbine or compressor applications the enthalpy and the kinetic energy are combined to form the total enthalpy h_o. Thus, (2.9) becomes

$$h_{o1} = h_{o2} + E \tag{2.10}$$

Compressors and pumps increase h_o so that $h_{o2} > h_{o1}$, and the energy transfer E is negative. On the other hand, turbines decrease h_o and E is positive. The work per unit mass calculated from (2.10), when divided by the gravitational acceleration g, becomes head H, which is the preferred form of work in pump or hydraulic turbine applications.

2.4   Momentum Equation

The moment of momentum equation is of particular interest in turbomachinery applications. In its general form it states that the sum of the moments of external forces on the fluid in a control volume equals the rate of increase of angular momentum in the control volume plus the net outflow of angular momentum from the control volume. Allen and Ditsworth (1972) give this in equation form as

$$\Sigma M = \frac{\partial}{\partial t} \int_{c.v.} \rho R \times V \, dv + \int_{c.s.} R \times V \rho V \cdot dS \tag{2.11}$$

where c.v. and c.s. refer to integration over the control volume or control surface.

Applying (2.11) to a general turbomachine, the control volume is the volume of fluid in the casing surrounding the rotor. Forces are applied to this fluid along the surface of the rotor, and the sum of their moments about some point on the rotor shaft is denoted by the term on the left side of (2.11). Assuming steady flow through the control volume, the first term on the right side of (2.11) is eliminated. Noting that the quantity $\rho V \cdot dS$ is the mass flow rate through an elemental area dS of the control surface, and that it has a positive sign at the outlet, a negative sign at the inlet, and is zero elsewhere, we have

$$\Sigma M = \int_{A_2} R \times V \, d\dot{m} - \int_{A_1} R \times V \, d\dot{m} \tag{2.12}$$

where A_1 and A_2 refer to the flow areas at the inlet and outlet, respectively.

Aligning the z axis with the rotor axis and taking the moment center at 0, as indicated in Figure 2.4, we evaluate the angular momentum per unit mass in (2.12) by the determinant

$$R \times V = \begin{vmatrix} i_r & i_u & k \\ r & O & z \\ V_r & V_u & V_a \end{vmatrix}$$

The magnitude of the z-component of the angular momentum per unit mass is

$$(R \times V)_z = \begin{vmatrix} 0 & 0 & 1 \\ r & O & z \\ V_r & V_u & V_a \end{vmatrix}$$

and the resulting scalar expression of moment about the z-axis is

$$M_z = \int_{A_2} V_u r \, d\dot{m} - \int_{A_1} V_u r \, d\dot{m} \tag{2.13}$$

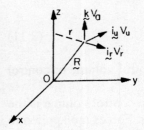

Figure 2.4 Velocity components.

Neglecting all forces other than those between the fluid and the rotor, we can say that the magnitude of the moment of forces M_z on the fluid in the control volume equals the negative of the torque T applied to the rotor shaft by the fluid. If we further assume constancy of the tangential component of the fluid velocity V_u and of the radial position r over the area A_1 or A_2, we can then write

$$T = \dot{m}(V_{u1}r_1 - V_{u2}r_2) \tag{2.14}$$

Turbomachine power is torque times the rotational speed N in radians per second. Thus power P can be expressed as

$$P = \dot{m}(V_{u1}U_1 - V_{u2}U_2) \tag{2.15}$$

Here we note that the blade speed U has been substituted for Nr. To obtain the energy transfer E per unit mass corresponding to that in (2.7) through (2.10), we simply divide (2.15) by the mass rate of flow $\dot{m}$. Thus, the energy transfer per unit mass from fluid to rotor, or vice versa, is given by

$$E = V_{u1}U_1 - V_{u2}U_2 \tag{2.16}$$

The latter relation is the Euler turbine equation, but it is applied to all types of turbomachines, including pumps and compressors.

2.5 Applications

Let us apply the above relations to a number of common turbomachines, namely, the axial-flow impulse turbine, the axial-flow compressor, the centrifugal pump, and the hydraulic turbine.

Impulse Turbine

Flow in the impulse turbine is generally in the axial direction, and the blade velocity is the same at the entrance and exit of the rotor. Figure 2.5 shows a typical blade cross section and the corresponding velocity diagram. Steam or hot gas leaves a nozzle with a velocity V_1 at a nozzle angle α, measured from the tangential direction, and enters the region between the blades with relative velocity W_1. Ideally, no pressure drop occurs in the blade passage, and the relative velocity W_2 is equal in magnitude to W_1. This is what is meant by the term *impulse turbine*, also called a zero-reaction turbine. The absolute velocity V_2 at the blade-passage exit is much reduced and is typically less than half of V_1. This energy, transferred from the fluid to the rotor, is found from (2.16) by making the substitutions $V_{u1} = V_1 \sin \alpha_1$, $V_{u2} = V_2 \sin \alpha_2$, and $U_1 = U_2 = U$. Thus,

$$E = U(V_1 \sin \alpha_1 - V_2 \sin \alpha_2) \tag{2.17}$$

The law of cosines applied to the two triangles in Figure 2.5 yields two equations, which when subtracted contain the right-hand side of (2.17). Substitution into (2.17) yields

$$E = \tfrac{1}{2}(V_1^2 - V_2^2 + W_2^2 - W_1^2) \tag{2.18}$$

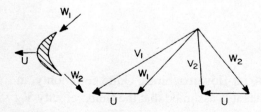

Figure 2.5 Velocity diagram for an impulse turbine.

Since $W_1 = W_2$, it is clear that E is really the difference in kinetic energy of the fluid, namely,

$$E = \tfrac{1}{2}(V_1^2 - V_2^2) \tag{2.19}$$

Maximizing energy transfer means minimizing V_2 or requiring that V_2 has only an axial component, i.e., $\alpha_2 = 0$.

The result of (2.19) is also obtainable from (2.9), if $h_1 = h_2$. Equal enthalpy implies no change of temperature and pressure in the flow, which agrees with the original assumption of zero reaction, i.e., no pressure drop in the rotor. If the degree of reaction of a turbine stage, denoted by R, is defined as the ratio of the enthalpy drop in the rotor to the enthalpy drop in the stator plus that in the rotor, namely,

$$R \equiv \frac{h_1 - h_2}{h_e - h_2} \tag{2.20}$$

where h_e is the enthalpy at the nozzle (stator) inlet, then we can write

$$h_e + \frac{V_e^2}{2} = h_2 + \frac{V_2^2}{2} + E \tag{2.21}$$

as the energy balance for the entire stage, and R becomes

$$R = \frac{V_2^2/2 - V_1^2/2 + E}{V_2^2/2 - V_e^2/2 + E} \tag{2.22}$$

Substituting (2.18) into (2.22) yields

$$R = \frac{W_2^2 - W_1^2}{V_1^2 - V_e^2 + W_2^2 - W_1^2} \tag{2.23}$$

which is generally applicable to axial-flow machines. Quite commonly, in the analysis of multistage machines it is assumed that the fluid velocity V_2 leaving the rotor is the same as that from the stage immediately upstream, i.e., $V_e = V_2$. The degree of reaction would then be expressed as

$$R = \frac{W_2^2 - W_1^2}{V_1^2 - V_2^2 + W_2^2 - W_1^2} \tag{2.24}$$

which we will utilize for axial-flow machines.

We have learned that the blade profile of the impulse turbine is designed to make $W_1 = W_2$. Clearly, (2.24) confirms the earlier assumption that $R = 0$.

Axial-Flow Compressor

An axial-flow compressor blade and velocity diagram are shown in Figure 2.6. The fluid is deflected only slightly by the moving blade compared with the turbine-blade deflection. Another difference is that the pressure rises in the flow direction both in the stator and in the rotor. Pressure rise is related to enthalpy rise, and the latter is dependent on the deflection of the fluid by the moving blade. A relationship is obtained by eliminating E between (2.9) and (2.18). This yields

$$h_2 - h_1 = \frac{1}{2}(W_1^2 - W_2^2) \tag{2.25}$$

which is interpreted as an enthalpy rise associated with a loss of relative kinetic energy in the blade passages. The associated pressure ratio is easily obtained from the enthalpy rise through the use of a polytropic exponent n. Using the perfect gas relation $\Delta h = C_p \Delta T$, we obtain

$$h_2 - h_1 = C_p T_1 \left[\left(\frac{P_2}{P_1} \right)^{(n-1)/n} - 1 \right] \tag{2.26}$$

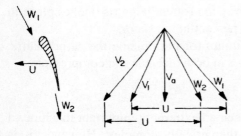

Figure 2.6 Velocity diagram for an axial-flow compressor.

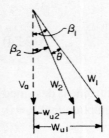

Figure 2.7 Tangential components of relative velocity.

from which the pressure ratio, and hence the pressure rise, may be determined. It is observed that pressure rise depends on the change of relative velocity, which is directly related to the compressor blade shape, i.e., to the angle of deflection of the fluid.

Energy transfer is also related to the deflection angle, since the application of (2.3) gives

$$E = U(V_{u1} - V_{u2}) = U(W_{u1} - W_{u2}) \tag{2.27}$$

Figure 2.7 shows that the difference in the tangential components of the relative velocity is proportional to the deflection angle $\theta = \beta_1 - \beta_2$.

A typical compressor velocity diagram is constructed by making $V_1 = W_2$ and $V_2 = W_1$. Referring to Figure 2.6, it is seen that the triangles would be symmetrical about the common altitude (V_a). Such symmetry, whether in a turbine or compressor diagram, results in $R = \frac{1}{2}$, as determined from (2.24). This condition is also termed a 50 percent reaction. Physically this means that 50 percent of the compression (or enthalpy rise) takes place in the rotor of the compressor and 50 percent in the stator.

This degree of reaction is optimum for minimizing the aerodynamic drag losses of rotor and stator blades in both turbines and compressors.

Centrifugal Pump

The centrifugal pump is and has been an extremely important machine to humans, and one would think it theoretically complex. However, it is extremely simple to analyze. It was discussed in Chapter 1 and is illustrated

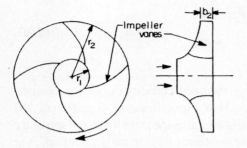

Figure 2.8 Centrifugal pump.

in Figure 2.8. The inner and outer radii r_1 and r_2 define the inlet and outlet of the control volume. Fluid, assumed incompressible, enters at station 1 with a purely radial velocity V_1, which implies that $V_{u1} = 0$. The impeller imparts angular momentum to the fluid so that it exits at station 2 with radial and tangential velocity components. Note that $U_2 > U_1$ since $r_2 > r_1$ and the angular speed is constant. The energy transfer E, or the head H ($gH = -E$), is calculated from (2.16) as

$$gH = -E = U_2 V_{u2} \qquad\qquad (2.28)$$

From Figure 2.9,

$$V_{u2} = U_2 - V_{m2}\tan\beta_2 \qquad\qquad (2.29)$$

The meridional component V_{m2} is the volume flow rate Q divided by the flow area $2\pi r_2 b_2$, and $U_2 = N r_2$. The head is thus expressed as

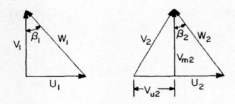

Figure 2.9 Velocity diagram for a centrifugal pump.

$$H = \frac{Nr_2[Nr_2 - (Q/2\pi r_2 b_2) \tan \beta_2]}{g}$$

(2.30)

where b_2 = impeller tip width.

An important performance curve, the head-capacity curve, for a centrifugal pump is constructed by plotting H as a function of Q. Equation (2.30) expresses this relationship analytically and provides an ideal head-capacity curve for comparison with actual curves. Since β_2 is usually about 65°, the theoretical relation indicates decreasing head with increasing flow rate, a situation realized in practice. This equation indicates that H goes up as the square of N, which also agrees with experience.

It is interesting to note that an actual pump impeller can be measured, and the measurements used to predict expected flow rate. Figure 2.9 shows that such a prediction can be easily made from a knowledge of β_1, N, r_1, and b_1, since

$$Q = 2\pi \, r_1^2 \, b_1 \, N \cot \beta_1$$

(2.31)

The enthalpy rise for a compressible fluid is determined from the thermodynamic equation

$$T \, ds = dh - \frac{dp}{\rho}$$

(2.32)

Equation (2.32), integrated for the ideal isentropic compression of a liquid (for which the density is assumed constant), is

$$h_2 - h_1 = \frac{p_2 - p_1}{\rho}$$

(2.33)

Substituting (2.33) into (2.9) gives

$$H = \frac{p_2 - p_1}{\rho g} + \frac{V_2^2 - V_1^2}{2g}$$

(2.34)

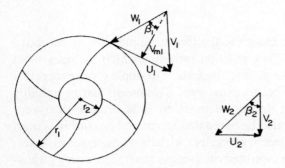

Figure 2.10 Hydraulic turbine.

The latter equation is useful in calculating the pressure rise across the pump impeller. Of course, the pressure can be raised further in the casing of the pump by reducing V_2 in a passage of increasing cross-sectional area, i.e., a diffuser.

Hydraulic Turbine

The radial-flow hydraulic turbine, depicted in Figure 2.10, is the reverse of the centrifugal pump. Stations 1 and 2 are reversed, and water enters at the larger radius r_1 from a stator which controls the angle at which the water enters the rotor inlet. Ideally, the absolute velocity V_2 at the exit is purely radial so that the energy transfer is simply

$$E = gH = U_1 V_{u1} \tag{2.35}$$

Since $V_{u1} = U_1 - V_{m1} \tan \beta_1$ and $V_{m1} = Q/A_1$, we can write the turbine head as

$$H = \frac{Nr_1 \{Nr_1 - [Q/(2\pi r_1 b_1)] \tan \beta_1\}}{g} \tag{2.36}$$

where b_1 = runner tip width. The flow rate Q is given, as in (2.31), by

$$Q = 2\pi r_2^2 b_2 N \cot \beta_2 \tag{2.37}$$

Further Examples

Similarities of other turbomachines to the four examples discussed above should be noted. The axial-flow reaction turbines, which includes most steam and gas turbines, are like the impulse turbine example given except that an expansion of the fluid also occurs in the rotor. This means that an enthalpy drop occurs there, and the degree of reaction R is greater than zero (typically $R = \frac{1}{2}$). It should also be noted that steam and gas turbines used to drive large loads, such as electric generators, include many stages in series, frequently with many rotors mounted on a single shaft. The energy transfer term for each rotor (stage) must be added to obtain the total work done per unit mass of steam or gas flowing. The turbine power is then obtained from the product of the total specific work and the mass flow of fluid.

The axial-flow compressor example indicated calculations for a single stage. Compressors usually involve many stages and the pressure ratios for each must be multipled to obtain the overall pressure ratio of the machine. In addition, the relations developed for the compressor stage would also apply to axial-flow blowers, fans, and pumps. The difference is that the isentropic enthalpy rise is calculated from (2.33) for the approximately incompressible flows usually assumed in these machines.

The centrifugal pump is geometrically similar to the centrifugal compressor, centrifugal blower and centrifugal fan. However, flow in the compressor must be modeled as compressible, and the pressure ratio should be calculated from an equation like (2.26). Usually, however, total properties p_o and T_o are used to formulate a working equation for the calculation of total pressure ratio in centrifugal compressor stages.

Starting with equation (2.10) we can write

$$| E | = c_p(T_{o2} - T_{o1}) \tag{2.38}$$

Since $V_{u1} = 0$ at the inlet of the centrifugal compressor, the Euler equation (2.16) simplifies to

$$| E | = V_{u2} U_2 \tag{2.39}$$

Eliminating E between (2.38) and (2.39) yields

$$c_p(T_{o2} - T_{o1}) = V_{u2} U_2 \tag{2.40}$$

Using a polytropic process to relate the end states we have

$$\frac{P_{o2}}{P_{o1}} = \left(\frac{T_{o2}}{T_{o1}}\right)^{n/(n-1)}$$

(2.41)

Combination of (2.40) and (2.41) results in the following working equation for the total pressure ratio of a centrifugal compressor stage

$$\frac{P_{o2}}{P_{o1}} = \left(1 + \frac{V_{u2}\, U_2}{c_p\, T_{o1}}\right)^{n/(n-1)}$$

(2.42)

Although (2.42) expresses the essential form for calculation of stage total pressure ratio, some additional refinement is required and will be added in Chapter 5.

As observed in Figure 2.8 fluid enters the eye of the centrifugal impeller axially, i.e., $V_{u1} = 0$, which means that equation (2.28) is also valid for the case where the pump or compressor vanes are extended into the eye of the impeller. However, it should be noted that, in this case, the cylindrical flow area $2\pi r_1 b_1$ used in equation (2.31) must be replaced by a circular or annular flow area. Moreover, the meridional velocity V_{m1}, which appears in (2.31) as $N r_1 \cot \beta_1$ must be replaced by the axial velocity V_1 entering the eye of the impeller.

A similar situation exists at the outlet of the hydraulic turbine rotor depicted in Figure 2.10, viz., the fluid can be made to exit axially, which implies that neither tangential nor radial velocity components are present, and the vanes can be extended into the exit plane of the rotor. For an assumed axial exit at station 2, $V_{u2} = 0$, as was assumed in the development of equation (2.35).

Likewise, equation (2.35) can be applied to the radial-inflow gas turbine, which is of the same geometry as the inward-flow hydraulic turbine. Hence, the vanes are usually extended into the exhaust plane, where there is no swirl of the exhaust gases and $V_{u2} = 0$. For the gas turbine the relative gas angle β_1 at the turbine inlet is zero, which implies that $V_{u1} = U_1$. These conditions define the so-called 90° (radial entry) IFR (Inward-Flow Radial) gas turbine, which is a design commonly employed in industry.

References

Allen, T., and R. L. Ditsworth. 1972. *Fluid Mechanics*. McGraw-Hill, New York.
Jones, J. B., and G. A. Hawkins. 1986. *Engineering Thermodynamics*. John
    Wiley & Sons, New York.

Problems

2.1. Construct the velocity diagram for an axial-flow gas turbine having a
      degree of reaction of 0.25 and minimum leaving kinetic energy ($V_2^2/2$).

2.2. Derive the relationship between torque and speed for an axial-flow
      impulse turbine.

2.3. Determine energy transfer E for an axial-flow turbine in terms of blade
      speed U when the degree of reaction is 0.5 and the leaving kinetic
      energy is minimal.

2.4. Repeat Problem 2.3 for the impulse turbine with minimal leaving
      kinetic energy.

2.5. Sketch the head-capacity curves for centrifugal pumps having β_2
      between 0° and 90°, equal to 0° and less than 0°.

2.6. Derive equation (2.18) using the law of cosines applied to the velocity
      triangles shown in Figure 2.5. Hint: Use the cosine of the nozzle angle α
      to form one equation and the angle opposite W_2 to form the other equation.

2.7. Consider a single-stage air turbine with an air flow rate of 2 kg/s. The
      gas turbine is a 90° IFR type. The relative velocity of the air entering
      the rotor is purely radially directed. The exhaust is purely axially
      directed. The rotor has an O.D. (outside diameter) of 0.3 m and the tip
      blade speed is 350 m/s. Find:
      a) the turbine power in kW
      b) the shaft speed in rpm
      c) the shaft torque in N-m
      Hint: The turbine is similar to the hydraulic turbine depicted in
      Figure 2.10 with $V_{u1} = U_1$ and $V_{u2} = 0$.

2.8. The impeller of a centrifugal pump has an O.D. of 0.30 m. Oil having
      a specific gravity of 0.81 enters the impeller at a rate of 63.0833 liters
      per second. The relative velocity at the impeller exit has no tangential
      component. The impeller rotates at 4800 rpm. Find:
      a) the impeller tip speed in m/s
      b) the energy transfer in J/kg
      c) the power input in kW

2.9. An axial-flow steam-turbine rotor receives steam at 20° to the tangential direction and exhausts it axially. The axial component of the steam velocity, which is assumed constant, is 0.7 times the blade velocity at the mean radius of the blade and has a magnitude of 450 fps. The mass flow rate of steam through the stage is 5.75 lb/s. For the stage at the mean radius, find:

    a) the relative velocity leaving the rotor in fps
    b) the relative velocity entering the rotor in fps
    c) the absolute velocity leaving the nozzles in fps
    d) the degree of reaction
    e) the energy transfer in ft-lb/slug
    f) the power produced in hp

2.10. An axial-flow gas-turbine stage has a degree of reaction of 0.5. The blade speed is 600 fps at the mean radius, and the mean radius is 1.5 ft. Gas enters the stage at a nozzle angle of 27°. The exhaust velocity is directed axially over the entire annulus, which has a throughflow area of 3.927 sq ft. The density of the exhaust gas is 0.135 lb/cu ft. The axial velocity component is constant through the rotor. Find:

    a) the mass flow rate of gas in lb/s
    b) the energy transfer in ft-lb/slug
    c) the rotational speed in rpm
    d) the power output in hp

2.11. Consider an axial-flow air-compressor stage. Air enters the stage from the atmosphere at a static temperature of 540°R. The blade speed at the mean radius is 1000 fps. The air deflection angle in both the rotor and the stator is 30°. The absolute velocity leaving the rotor makes an angle of 60° with the axial direction. The degree of reaction of the stage is 50 percent. Assume that the stage is a repeating stage, i.e., $V_1 = V_3$, that the axial component is the same at all three stations and that the compression is isentropic. Find the stage (static) pressure ratio.

2.12. An axial-flow gas-turbine stage is to be designed for 50 percent reaction and minimum exhaust gas kinetic energy. The rotor is to turn at 3600 rpm with a mean blade velocity of 1000 fps. The nozzle blades are set to produce a gas nozzle angle of 24 degrees. Assume that $C_p = 0.27$ Btu/lb-°R and that the stage exhausts to the atmosphere at 14.1 psia and 900°F. Find:

    a) the total temperature drop in the stage
    b) the energy transfer in Btu/lb
    c) the power produced in horsepower for a blade length of 8 in.

2.13. Solve for the stage horsepower of an air turbine with $V_{u1} = U$ and $V_{u2} = 0$. The mean blade velocity is 1200 fps, and the mass rate of flow of air through the stage is 50 lb/s. The nozzle angle α is 15°. For an annular throughflow area A_2 of 4.99 sq ft, find the density ρ_2 of the exhaust gas in lb/cu ft.

2.14. For the axial-flow compressor show that degree of reaction $R = -W_{um}/U$, where $W_{um} = (W_{u1} + W_{u2})/2$.

2.15. Show that the degree of reaction R is also given by $R = -W_{um}/U$ for axial flow turbines.

2.16. Consider an axial-flow gas turbine with $R = \frac{1}{2}$. The exhaust velocity V_2 makes an angle of 10° with the axial direction. The decrease of tangential velocity in the rotor is $\Delta V_u = 1250$ fps. The blade speed $U = 1100$ fps. Find:
    a) the energy transfer E
    b) the axial component of velocity V_a
    c) the nozzle exit velocity V
    d) the nozzle angle α

2.17. A centrifugal pump handles 2400 gallons per minute of water (density = 62.4 lb/cu ft). The impeller diameter D_2 is 19 inches, the impeller tip width b_2 is 1.89 inches, the rotational speed is 870 rpm, and the head H is 80 feet. Find:
    a) the impeller tip speed U_2 in fps
    b) the meridional velocity W_{m2} in fps
    c) the tangential velocity component V_{u2} in fps
    d) the relative fluid angle β_2 in degrees

2.18. A centrifugal water pump with backward-curved vanes runs at 1500 rpm and delivers 0.6688 cu ft/s. The tip width of the vanes is $b_2 = 0.5$ inch. The tip diameter $D_2 = 0.5106$ ft. The head H = 28 feet. The density of water is 62.4 lb/cu ft. Find:
    a) the tangential velocity component at the impeller tip V_{u2} in fps
    b) the tip speed U_2 in fps
    c) the angle β_2 between W_2 and U_2 in degrees.

2.19. An axial-flow gas turbine stage produces 1000 hp with a gas flow rate of 10 lb/sec. The degree of reaction is 50 percent. The exhaust velocity $V_2 = 600$ fps and is directed axially. Find the blade speed U and the nozzle angle α.

2.20. The nozzle angle (angle between V_1 and U) is 25° in an axial-flow gas turbine. The through-flow velocity $V_a = 400$ fps, and the mean blade speed $U = 800$ fps. The velocity diagram is symmetrical ($W_1 = V_2$). Find:

a) the relative fluid angles β_1 and β_2
b) the energy transfer E
c) $h_1 - h_2$

2.21. Use the through-flow and blade velocity data from Problem 2.20 to construct the velocity diagram for an impulse turbine with zero swirl in the exhaust gas ($V_{u2} = 0$). Find:

a) the nozzle angle α
b) the energy transfer
c) $h_1 - h_2$

Symbols for Chapter 2

A	area of flowing stream at control surface
b_1	width of vane at $r = r_1$ in pump or turbine rotor
b_2	width of vane at $r = r_2$ in pump or turbine rotor
c_p	specific heat of gas at constant pressure
c_v	specific heat of gas at constant volume
e	specific internal energy of fluid
E	energy transfer = w
H	head = $-E/g$
h	specific enthalpy of the fluid
h_o	specific total (stagnation) enthalpy of the fluid
i_r	unit vector in radial direction
i_u	unit vector in tangential direction
k	unit vector in axial direction (z-direction)
$\dot{m}$	mass flow rate of fluid
N	rotor speed
n	polytropic exponent ($n = \gamma$ for isentropic processes)
p	fluid pressure
P	power to or from rotor
q	heat transfer per unit mass of flowing fluid
r	radial coordinate perpendicular to axis of rotation
R	position vector; origin on axis of rotation
R	degree of reaction
T	absolute temperature
U	blade speed
V	absolute velocity
V_a	component of V in axial direction

V_m component(meridional) of V perpendicular to control surface

V_r component of V in radial direction

V_u component of V in tangential direction

w   specific work done on or by the fluid

W   velocity relative to moving blade

W_u tangential component of W

z   altitude above an arbitrary plane in a direction opposite to g

α nozzle angle; angle between V_1 and U

α_1 angle between V_1 and V_{m1} ($V_a = V_{m1}$ in axial-flow machines)

α_2 angle between V_2 and V_{m2} ($V_a = V_{m2}$ in axial-flow machines)

β_1 angle between W_1 and V_{m1} ($V_{m1} = V_a$ in axial-flow machines; $V_{m1} = V_{r1}$ in radial-flow machines)

β_2 angle between W_2 and V_{m2} ($V_{m2} = V_a$ in axial-flow machines; $V_{m2} = V_{r2}$ in radial-flow machines)

γ ratio of specific heats = c_p/c_v

ρ fluid density

θ angle between W_1 and W_2 (axial-flow compressor)

3 Dimensionless Quantities

3.1 Introduction

In plotting the results of turbomachinery tests and in the analysis of performance characteristics, it is useful to use dimensionless groups of variables. Appropriate groups of variables are found by application of a dimensional methodology (dimensional analysis), and it is known from the Buckingham pi theorem that the dimensionless groups so formed have a functional relationship, although the nature of the relationship is frequently unknown except by experimentation. An important benefit of dimensional analysis is that the results of model studies so analyzed and plotted may then be used to predict full-scale performance. This is important to reduce the cost of the development of turbomachines. It is also useful to use in the analysis of data from full-sized machines, when it is desired to predict the performance of other full-sized machines of a different size than those tested, or to operate under different conditions.

3.2 Turbomachine Variables

The important variables in turbomachine performance are shown in Table 3.1. The Buckingham pi theorem applied to the four variables and two dimensions indicates that two dimensionless groups can be formed. Of the several possible groups that can be formed, the most useful combinations of variables are the flow coefficient φ defined by

Table 3.1 Primary Turbomachine Variables

Variable	Symbol	Dimensions
Head or energy transfer	gH (E)	L^2/T^2
Volume flow rate	Q	L^3/T
Angular speed	N	$1/T$
Rotor diameter	D	L

$$\varphi = \frac{Q}{ND^3} \qquad \textit{Flow Coefficient} \tag{3.1}$$

and the head coefficient ψ defined as

$$\psi = \frac{gH}{N^2D^2} \qquad \textit{head Coefficient} \tag{3.2}$$

where $|E| = gH$ has been used in the analysis.

Three other groups are also used extensively by engineers. However, they may be easily derived from φ and ψ. Specific speed N_s is formed in the following way:

$$N_s = \frac{\varphi^{1/2}}{\psi^{3/4}} = \frac{NQ^{1/2}}{(gH)^{3/4}} \qquad \textit{Specific Speed} \tag{3.3}$$

As φ and ψ are nondimensional flow rate and head, so N_s is nondimensional speed. In fact, if Q and H are unity, we observe that $N_s = N$. The other particularly useful dimensionless groups are the specific diameter D_s and the power coefficient C_p as defined below:

$$D_s = \frac{\psi^{1/4}}{\varphi^{1/2}} = \frac{D(gH)^{1/4}}{Q^{1/2}} \qquad \textit{Specific Diameter} \tag{3.4}$$

$$C_p = \varphi\psi = \frac{P}{\rho N^3 D^5} \qquad \textit{Power Coefficient} \tag{3.5}$$

where power $P = \rho QgH$.

Recall that four variables are assumed to be of primary importance. If additional variables are added to the list in Table 3.1, then a new dimensionless group can be formed which will contain each, i.e., an additional group for each new variable. For example, if kinematic viscosity v is added, we have the Reynolds number Re, defined as

$$Re = \frac{ND^2}{v} \quad \text{Reynolds number}$$

(3.6)

Another example is the inlet fluid temperature T_1, or the inlet specific enthalpy h_1. Since the latter quantity contains the square of the acoustic speed in gases, we would expect the Mach number M to emerge as the appropriate dimensionless group, namely,

$$M = \frac{ND}{\sqrt{h_1 (\gamma - 1)}} \quad \text{Mach number}$$

(3.7)

This is an important variable in turbomachines involving high-speed flow of gases.

If the number of variables is increased to seven by adding the inlet pressure P_1, the nature of the dimensionless groups is changed but not the number. This is because pressure involves the force dimension, which is not present in the others. Thus, the number of groups remains four. However, since density ρ_1 can be introduced as a combination of p_1 and T_1 (from h_1), it appears in the groups; for example, in combination with Q as the mass flow rate $\dot{m}$. In this case the mass flow coefficient can be written as

$$C_M = \frac{\dot{m}}{(p_1\rho_1)^{1/2} D^2} \quad \text{Mass flow Coefficient}$$

(3.8)

Other forms of the head coefficient are the ratio of outlet pressure to inlet pressure p_2/p_1, or the ratio of outlet to inlet temperature T_2/T_1. Clearly, these ratios are equivalent, since head is proportional to enthalpy difference, which in turn is proportional to temperature difference in gas-flow machines and pressure difference in incompressible-flow machines. In gas-flow turbomachines either p_2/p_1 or T_2/T_1 could be used, since the two are related through isentropic or polytropic process relations.

Efficiency η has many specialized definitions, but is, in general, output power divided by input power. It, too, can be included in the list of variables, and since it is already dimensionless, it is also included in the list of dimensionless groups.

3.3 Similitude

Flow similarity occurs in turbomachines when geometric, kinematic, and dynamic similarity exist between a model (i.e., a small-sized turbomachine) and its larger prototype. Thus, ratios of dimensions of corresponding parts are the same throughout. Velocity triangles at corresponding points in the flow fields are also similar, as are ratios of forces acting on the fluid elements. Similar velocity triangles, for example, imply equal flow coefficients:

$$\left(\frac{Q}{ND^3}\right)_p = \left(\frac{Q}{ND^3}\right)_m \tag{3.9}$$

In contrast, similar force triangles are equivalent to equal head coefficients:

$$\left(\frac{gH}{N^2D^2}\right)_p = \left(\frac{gH}{N^2D^2}\right)_m \tag{3.10}$$

The equality of dimensionless groups resulting from similitude has important practical consequences. It allows a most compact presentation of graphical results. One example of this is seen from a consideration of head-capacity curves for centrifugal pumps, which typically appear as shown in Figure 3.1. A separate curve is needed for each shaft speed when plotting the primary variables. On the other hand, if head coefficient is plotted against flow coefficient, the curves collapse to a single curve, and a single relationship exists between ψ and φ, regardless of speed.

The so-called pump laws also follow from the similarity conditions expressed by (3.9) and (3.10). When it is desired to know how a given pump will perform at another speed when its performance at one speed is known, we simply cancel the D^3 factors in (3.9) and find

$$\frac{Q_1}{N_1} = \frac{Q_2}{N_2} \tag{3.11}$$

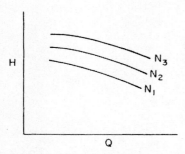

Figure 3.1 Head-capacity curves.

which is a pump law; it implies that capacity Q varies directly with speed N. In a similar manner, we see from (3.10) that the head H or pressure rise is governed by

$$\frac{H_1}{N_1^2} = \frac{H_2}{N_2^2}$$

(3.12)

i.e., head varies as the square of the speed. Power is the product of Q and H, and the third pump law states that

$$\frac{P_1}{N_1^3} = \frac{P_2}{N_2^3}$$

(3.13)

Laws for scaling up or down, i.e., varying diameter D while keeping the speed constant, follow in a similar manner after canceling the factors containing N. Thus we find

$$\frac{Q_1}{D_1^3} = \frac{Q_2}{D_2^3}$$

(3.14)

$$\frac{H_1}{D_1^2} = \frac{H_2}{D_2^2}$$

(3.15)

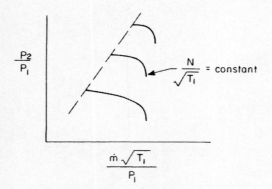

Figure 3.2 Compressor map.

$$\frac{P_1}{D_1^5} = \frac{P_2}{D_2^5}$$

(3.16)

Performance curves are frequently plotted from dimensionless or quasi-dimensionless groups. Compressor maps, for example, are usually presented in graphs of the form shown in Figure 3.2. The abscissa is determined from (3.8) by dropping the diameter, since it is not a variable in the performance of a single machine, and by using the same gas constant. Similarly, the parameter $N/\sqrt{T_1}$ is a variation of the machine Mach number formed from (3.7) by eliminating specific heat and the rotor diameter, which are both constants for a given map.

Table 3.2 Specific Speeds

Turbomachine	Specific speed range
Pelton wheel	0.03–0.3
Francis turbine	0.3–2.0
Kaplan turbine	2.0–5.0
Centrifugal pumps	0.2–2.5
Axial-flow pumps	2.5–5.5
Centrifugal compressors	0.5–2.0
Axial-flow turbines	0.4–2.0
Axial-flow compressors	1.5–20.0

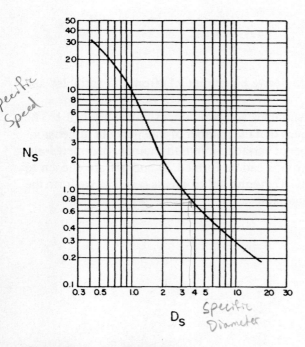

Figure 3.3 Cordier diagram.

Source: G.T. Csanady, *Theory of Turbomachines.* Copyright 1964 by McGraw-Hill, Inc. Used with the permission of McGraw-Hill Book Company.

Besides their use in performance curves, dimensionless groups are useful in design and machine selection. For example, specific speed is commonly used to indicate the type of machine appropriate to a given service. Table 3.2 gives ranges of specific speeds corresponding to efficient operation of the turbomachines listed. Sizes of turbomachines required for a given service are also determinable from specific speed-diameter plots of the type shown in Figure 3.3. This correlation, developed by Csanady (1964), of the optimum specific speeds of various machines as a function of specific diameter is useful in determining an appropriate size for a given set of operating conditions. To enter the diagram, called the Cordier diagram, a specific speed can be selected from Table 3.2. The rotor diameter can be determined from the specific diameter found from the Cordier diagram. A machine so selected or designed would be expected to have high efficiency.

3.4   Examples

Example Problem 3.1

Use dimensional analysis to derive equation (3.1) from the variables and dimensions in Table 3.1.

Solution: There are four variables and two dimensions in Table 3.1. The Buckingham Pi theorem states that the number of independent dimensionless groups equals the number of variables minus the number of dimensions. Since two dimensionless groups can be formed, Q and gH are chosen to serve as nuclei for the groups. Denoting the groups as φ and ψ, we form the following arrangements of variables:

$$\varphi = QN^aD^b$$

and

$$\psi = gHN^cD^d$$

where the exponents of N and D are to be determined.

Since φ is dimensionless, we can write the φ -equation dimensionally as

$$L^oT^o = L^3T^{-1}T^{-a}L^b$$

Equating the exponents of L we have

$$0 = 3 + b$$

The equation for the exponents of T is

$$0 = -1 - a$$

Hence, $a = -1$ and $b = -3$, and the result is

$$\varphi = QN^{-1}D^{-3}$$

Example Problem 3.2

Calculate the dimensionless and the dimensional values of specific speed
N_s for a centrifugal water pump whose design-point performance is the
following:

$$Q = 2400 \text{ gpm}$$
$$H = 70 \text{ feet}$$
$$N = 870 \text{ rpm}$$

Solution: The first step is to convert each quantity into consistent units.
Use the conversion factors from Table 1 in Appendix A.

$$Q = \frac{2400}{448.8} = 5.3476 \text{ cfs}$$

$$\hookrightarrow = 1 \frac{ft^3}{s} = 448.8 \frac{gallons}{m}$$

$$E = gH = (32.174)(70) = 2252 \text{ ft-lb}_F/\text{slug}$$

$$N = \frac{870}{9.5493} = 91.106 \text{ rad/sec}$$

$$\frac{60}{2\pi} = 9.5493$$

The dimensionless specific speed is

$$N_s = \frac{\overset{N}{91.106}\,\overset{Q}{(5.3476)^{1/2}}}{\underset{gH}{(2252)^{3/4}}} = 0.644$$

The dimensional specific speed is

$$N_s = \frac{870(2400)^{1/2}}{(70)^{3/4}} = 1761$$

Example Problem 3.3

Use the relations for fluid friction losses in pipes to derive a scaling law for
the efficiency of hydraulic turbines. The scaling laws already derived are
for Q, H, and P and are given by equations (3.14)–(3.16). Such laws
are based on the principle of geometric, kinematic, and dynamic similarity.
Assume this similarity between model and prototype in the present derivation.
Solution: Refer to the Moody diagram in a fluid mechanics text, e.g., Fox
and McDonald (1985). The hydraulic loss H_L is calculated from a friction

factor f, which can be read from the Moody diagram. The hydraulic loss in a rotor flow passage is given by

$$H_L = \frac{fL_R W^2}{2gD_R} \tag{3.17}$$

Defining efficiency as output over input, we write

$$\eta = \frac{g(H - H_L)}{gH} \tag{3.18}$$

where gH is the mechanical energy given up by the fluid during its passage through the turbine. The numerator of (3.18) represents the mechanical energy extracted by the turbine rotor. The dimensionless loss is given by

$$1 - \eta = \frac{H_L}{H} = \frac{fL_R W^2 N^2 D^2}{8D_R U^2 gH} \tag{3.19}$$

Similarity between model and prototype implies equality of L_R/D_R, W/U and $gH/(N^2D^2)$. Thus we find

$$\frac{1 - \eta_p}{1 - \eta_m} = \frac{f_p}{f_m} \tag{3.20}$$

where subscripts p and m refer to prototype and model, respectively.

In turbomachines the Reynolds number is typically so high that only the relative roughness of the passage walls affects the friction factor. Because of similarity D/D_R is also constant. If roughness height is assumed invariant, then friction factor is inversely related to machine diameter, and

$$\frac{1 - \eta_p}{1 - \eta_m} = \left(\frac{D_m}{D_p}\right)^n \tag{3.21}$$

where n is an experimentally determined exponent.

References

Csanady, G. T. 1964. *Theory of Turbomachines*. McGraw-Hill, New York.
Fox, R. W., and A. T. McDonald. 1985. *Introduction to Fluid Mechanics*.
    John Wiley & Sons, New York.

Problems

3.1. Derive expressions for specific speed, specific diameter, and power coefficient by combining flow and head coefficient. Show that each is dimensionless.

3.2. Determine the model-to-prototype diameter ratio for a water turbine which will produce 30,000 hp at 100 rpm with a head of 50 ft, while the model will produce 55 hp under a head of 15 ft. Hint: Use the result of Problem 3.19.

3.3. Determine the speed and flow rate of the model turbine in Problem 3.2. Assume that the turbine efficiency η is 0.90. Hint: Use the turbine power relation $P = \eta \rho g Q H$.

3.4. A centrifugal fan is to be compared with a larger, geometrically similar fan. The smaller fan delivers 500 ft^3/min of air at standard conditions with a pressure rise (head) of 2 in. of water. The smaller fan runs at 1800 rpm, while the larger one operates at 1400 rpm while producing the same head. Determine the diameter ratio of the two fans and the flow rate of the larger.

3.5. Use the Cordier diagram to estimate the rotor diameter of a pump which, while running at 1000 rpm at a head of 30 ft, will deliver 4500 gal/min. Would you recommend a centrifugal or an axial-flow pump for this service?

3.6. Use dimensional analysis to derive equation (3.2) from the variables and dimensions in Table 3.1.

3.7. Derive equation (3.6) by forming a new group, Reynolds number Re, in which the kinematic viscosity appears, along with N and D. The Reynolds number so formed is called the Machine Reynolds number. Hint: Start with $Re = v^{-1}N^a D^b$. Note that the dimensions of v are L^2/T.

3.8. Derive equation (3.7) by introducing the specific enthalpy h_1, the enthalpy at the inlet of a machine in which the flow is compressible, as an additional variable. Note that the dimensions of specific enthalpy are FL/M, i.e., energy dimensions divided by mass dimensions. Additionally, Newtons Second Law can be written dimensionally as

$F = ML/T^2$. Hence, the dimensions of h are L^2/T^2, and no new dimensions are introduced. Hint: Form the new dimensionless group, Mach number, by writing $M = h^{-\frac{1}{2}}N^aD^b$.

3.9. Show that the quantity h_1 in equation (3.7) is the same as the square of the acoustic speed at temperature T_1 divided by $\gamma - 1$, where γ is the ratio of specific heats.

3.10. Show that the quantity ND, which appears in equations (3.6) and (3.7), is proportional to the blade speed at the tip of a turbomachine rotor.

3.11. Show that the flow coefficient φ defined in equation (3.1) is proportional to V_a/U in an axial-flow turbomachine. The latter form is used to define the flow coefficient in Chapters 6 and 8.

3.12. Show that the flow coefficient φ defined in equation (3.1) is proportional to W_{m2}/U_2 in centrifugal pumps and compressors. The latter definition is used to define the flow coefficient in Chapters 4 and 5. Hint: Use the fact that b_2 is proportional to D_2 in centrifugal machines.

3.13. Show that the mass flow coefficient C_M defined by equation (3.8) can be derived from the product of φ and M, i.e., from equations (3.1) and (3.7). Hint: Divide φM by $[(\gamma - 1)/\gamma]^{\frac{1}{2}}$, and then multiply by ρ_1/ρ_1 to obtain equation (3.8).

3.14. Specific speed can be calculated as a dimensional quantity, e.g., as used in Table 2 in Appendix A, or in dimensionless form, as used to construct the Cordier diagram in Figure 3.3. In dimensional form, its units are $(rpm)(gpm)^{1/2}(ft)^{-3/4}$. Note that H is used in lieu of gH in calculating the dimensional form of N_s. Convert the dimensional specific speed $N_s = 1000$ into its corresponding dimensionless value.

3.15. If the velocity triangles of a centrifugal pump are similar to those of the model pump, show that the flow coefficients are equal for model and prototype. Hint: Use the results of Problem 3.12.

3.16. If the velocity triangles of an axial-flow compressor are similar to those of the model compressor, show that equality of the flow coefficients is a consequence. Hint: Use the results of Problem 3.11.

3.17. If the velocity triangles of an axial-flow compressor are similar to those of the model compressor, show that equality of head coefficients is a consequence. Hint: First show that $gH/(N^2D^2) = |E|/(4U^2)$, and then use equation (2.27).

3.18. Use the Cordier diagram (Figure 3.3) to determine a value of impeller diameter D_2 which is suitable for a centrifugal pump whose operating conditions are those given in Example Problem 3.2. Hint: Find the specific diameter D_s from the Cordier diagram using the previously calculated N_s. Use equation (3.4) to solve for D, which is the same as D_2 for the pump. Note: The actual pump with the same capacity and head has an impeller diameter of 19 inches.

3.19. Derive the relation which states that $NP^{1/2}/H^{5/4}$ is the same for model and prototype. Hint: Divide the square root of power coefficient C_p by the head coefficient ψ to the $5/4$ power. The dimensionless group formed involves N, P, g, H, and ρ. Equate these groups for model and prototype. Assume the same fluid density for model and prototype.

3.20. Derive a scaling law for the efficiency of centrifugal pumps. Use the method of Example Problem 3.3, but note that pump efficiency is defined by

$$\eta = \frac{gH}{g(H + H_L)}$$

where gH represents the mechanical energy rise of the fluid passing through the pump, and the denominator represents the mechanical energy transferred from the impeller to the fluid.

3.21. Use values from Table 2 in Appendix A to estimate the efficiency of the centrifugal pump described in Example Problem 3.2. Determine the hydraulic loss H_L from the definition of pump efficiency given in Problem 3.20.

3.22. Show that Q is proportional to D^3 for similar machines operating at the same specific speed N_s and the same rotational speed N. Hint: Use equation (3.3). Note that head and flow coefficients have the same values for model and prototype.

3.23. Estimate the empirical exponent in the scaling law derived in Problem 3.20 for centrifugal pumps with $N_s = 500$ and Q between 30 and 1000 gpm. Use data from Table 2 in Appendix A, and assume that Q is proportional to D^3. Hint: Plot $(1 - \eta)/\eta$ as a function of Q on log-log graph paper to determine the slope of the resulting straight line.

3.24. If the pump described in Example Problem 3.2 were to be operated at 1200 rpm, what capacity Q, head H, and power would result?

3.25. A model of the centrifugal pump analyzed in Example Problem 3.2 is to be constructed for laboratory study. Assuming complete similarity of model and prototype, that the capacity of the model is reduced to 150 gpm and that the same rotational speed is to be used for model and prototype, find the impeller diameter of the model, the head produced by the model pump and the model pump efficiency.

3.26. Use the Cordier diagram to test the manufacturer's claim that a centrifugal pump with a rotor diameter of 6 inches can supply water at the rate of 10 cfs and produce a head of 30 feet at a rotational speed of 980 rpm.

3.27. Examine the claim of pump performance stated in Problem 3.26 by calculating the maximum possible head with equation (2.30). Assume that radial vanes are used, so that $\beta_2 = 0°$. Is the 30-foot head possible even under these idealized conditions?

3.28. Use equation (2.30) to calculate the maximum possible head, i.e., the ideal head, for the pump described in Example Problem 3.2. Assume $D_2 = 19$ inches, $b_2 = 0.10D_2$ and $\beta_2 = 65°$.

Symbols for Chapter 3

a	exponent
b	exponent
b_2	width of vane at $r = r_2$ in pump or turbine rotor
c	exponent
C_M	mass flow coefficient
C_p	power coefficient
d	exponent
D	tip (radial-flow) or mean (axial-flow) rotor diameter
D_2	rotor tip diameter
D_s	specific diameter
D_R	diameter of rotor flow passage
E	energy transfer between rotor and fluid
F	force dimension
f	friction factor
g	gravitational acceleration
gH	loss (turbine) or gain (pump) of fluid mechanical energy during passage through the turbomachine
H	head = $|E|/g$
H_L	hydraulic loss in rotor passage
h	specific enthalpy of the fluid
h_1	specific enthalpy of the fluid at inlet
k_R	average height of surface roughness

k_R/D_R relative roughness of rotor passage
L   length dimension
L_R length of rotor passage
M   mass dimension
M   Mach number
$\dot{m}$ mass flow rate of fluid
N   rotor speed
N_s specific speed
n   empirical exponent in efficiency scaling law
P_1 fluid pressure at inlet
P   power to or from rotor
Q   capacity = flow rate
Re   Reynolds number
T   time dimension
T_1 absolute temperature at inlet
U   blade or runner speed
U_2 rotor tip speed
V_a axial component of absolute velocity V
W_{m2} meridional component of W at rotor exit
W   average velocity relative to moving rotor flow passage
β_2 angle between W_2 and W_{m2}
γ ratio of specific heats
ν kinematic viscosity
ρ fluid density
φ flow coefficient
ψ head coefficient
[=]   dimensionally equal to

4 Centrifugal Pumps and Fans

4.1 Introduction

Rotors, known more commonly as impellers, of centrifugal pumps, blowers, and fans are designed to transfer energy to a moving fluid that is considered incompressible. Fans and blowers usually consist of a single impeller spinning within an enclosure, known as the casing. Pumps, on the other hand, may be designed to have several impellers mounted on the same shaft, and the fluid discharging from one is conducted to the inlet of the neighboring rotor, thus making the overall pressure rise of the pump the sum of the individual-stage pressure rises. The individual impellers are designed to look, in cross section, somewhat like that shown in Figure 1.1.

An end view of the impeller is shown in Figure 4.1. The vanes shown are curved backwards making the angles β_1 and β_2 with tangents to the circles at radii r_1 and r_2, respectively. Ideally, the relative velocity W_2 leaves the vane at the outer edge of the impeller at the blade angle β_2.

Figure 4.2 shows velocity diagrams at the inlet and outlet of the vane passages. For the design-point operation the relative velocity W_1 is approximately aligned with the tangent to the vane surface at angle β_1. The absolute velocity V_1 at the inlet is shown entering with no whirl. Thus $V_{u1} = 0$ and $V_{m1} = V_1$. The ideal or virtual head H_i, which is the ideal energy transfer per unit mass for perfect guidance by the vanes, is given by

$$H_i = \frac{U_2 V_{u2}}{g} \qquad (4.1)$$

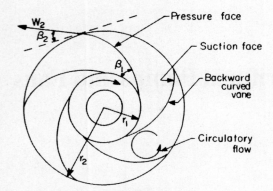

Figure 4.1 Pump impeller.

Equation (4.1) was derived earlier as (2.28). The ideal head H_i is higher than that found in practice. Reasons for this disparity and methods for correction will be given in subsequent sections.

4.2   Impeller Flow

Figure 4.1 shows an impeller rotating in the clockwise direction. Fluid next to the pressure face of the vane is forced to rotate at blade speed. Motion in a purely circular path at radius r implies a net pressure force directed radially in inward, so that the net pressure force A dp on a differential element of cross-sectional area A balances the centrifugal force $(\rho\ A\ dr)N_2 r$; thus the radial pressure gradient is

$$\frac{dp}{dr} = \frac{\rho U^2}{r} \tag{4.2}$$

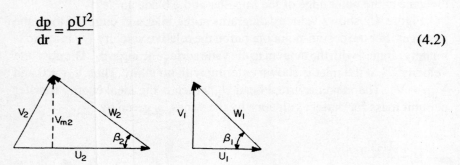

Figure 4.2 Velocity diagrams at inlet and outlet.

Since the fluid does not follow the impeller as in solid body rotation, but instead tends to remain stationary relative to the ground, a resultant outward flow along the vane with an accompanying adverse pressure gradient occurs. However, the magnitude of the pressure rise across the rotor is less than that indicated by integration of (4.2); i.e., less than

$$P_2 - P_1 = \frac{\rho(U_2^2 - U_1^2)}{2} \tag{4.3}$$

A better estimate of the pressure rise is obtained from an equation formed from (2.28) and (2.34):

$$P_2 - P_1 = \rho U_2 V_{u2} - \frac{V_2^2 - V_1^2}{2} \rho \tag{4.4}$$

Applying the law of trigonometry to the diagrams of Figure 4.2 yields the relations

$$U_2 V_{u2} = \frac{U_2^2 + V_2^2 - W_2^2}{2} \tag{4.5}$$

and

$$V_1^2 = W_1^2 - U_1^2 \tag{4.6}$$

Combining (4.4), (4.5), and (4.6) results in the pressure rise expression

$$P_2 - P_1 = \frac{\rho(U_2^2 - U_1^2 + W_1^2 - W_2^2)}{2} \tag{4.7}$$

Although the static pressure at the inlet and outlet of the impeller is expected to be uniform across the opening between the vanes, pressures on the two sides of a vane are expected to be different. As the fluid moves radially outward, its angular momentum per unit mass $V_u r$ is clearly increased. This means that a moment of some force has been applied to the control volume considered. The source of such a force is obviously a pressure difference between any two points on opposite sides of the control volume at the same radial distance from the axis of rotation. The azimuthal force resulting from this pressure difference is the so-called Coriolis force, of magnitude $|2N \times W|$. This force is applied to the impeller at the pressure face and the suction face of the vane. Equation (4.7) applied between the inlet and some intermediate radius less than r_2 implies that the greater pressure rise on the pressure face is accompanied by a lower relative velocity W on that face. Conversely, a higher relative velocity at the suction face is indicated. Figure 4.1 shows a circulatory flow which is radially inward on the pressure face and radially outward on the suction face, and this is superposed on the main flow, which is radially outward. The difference in pressure rise on the two sides of the passage between vanes implies a separation, or backflow, region near the outer end of the suction face. The latter implies a flow deflection away from the suction face near the exit of the passage. The change in V_{u2} associated with this flow deflection is known as slip.

The ratio of the actual V_{u2} to the ideal V_{u2} is usually known as the slip coefficient μ_s. Since the slip depends on the circulation, and the circulation is clearly dependent on the geometry of the flow passage, a theoretical relationship expressing μ_s as a function of the number of blades n_B and exit angle β_2 is not surprising. Shepherd (1956) has given such a relation:

$$\mu_s = 1 - \frac{\pi U_2 \sin \beta_2}{V_{u2} n_B}$$

(4.8)

which is derived in Appendix B. For a finite number of vanes, the velocity diagram of Figure 4.2 must be modified to reflect the effect of slip; this effect is illustrated in Figure B2 in Appendix B. The actual tangential component of V_2 is denoted by $V_{u2'}$, which replaces the component V_{u2}, i.e., that corresponding to perfect guidance by the vanes. The fluid angle for perfect guidance is β_2 and is the same as the vane angle. With a finite number of vanes and the accompanying slip, the actual fluid angle is different from the vane angle and is denoted by $\beta_{2'}$. The energy transfer with a finite number of vanes is given by $V_{u2'} U_2$, and the corresponding input head H_{in} is calculated from

$$H_{in} = U_2 \frac{V_{u2'}}{g}$$

(4.9)

4.3 Efficiency

Flow in the impeller or casing passages is accompanied by frictional losses which are proportional to the square of the flow velocity relative to the passage walls. All losses result in a conversion of mechanical energy into thermal (internal) energy. *Wall friction* effects this transfer through direct dissipation by viscous forces and by turbulence generation which culminates in viscous dissipation within the small eddies. *Secondary flow* losses occur in regions of flow separation, where circulation is maintained by the external flow, and in curved flow passages, where it is maintained by centrifugal effects.

The steady flow energy equation is applied to a control volume which is bounded by the pump casing and the suction and discharge flanges, as is depicted in Figure 4.3. The enthalpy, kinetic energy, and potential energy are changed by the work input gH_{in}, and the balance of these energies is expressed by

$$h_s + \frac{V_s^2}{2} + z_s g + gH_{in} = h_d + \frac{V_d^2}{2} + z_d g \tag{4.10}$$

where the subscripts s and d refer to properties at the suction and discharge flanges of the pump casing. Enthalpy can be written in terms of internal energy and flow work, so that the work input becomes

$$gH_{in} = \frac{V_d^2 - V_s^2}{2} + (z_d - z_s)g + \frac{p_d - p_s}{\rho} + e_d - e_s \tag{4.11}$$

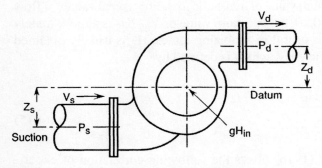

Figure 4.3 Centrifugal pump with piping.

The hydraulic loss gH_L is the loss of mechanical energy or the gain of internal energy per unit mass of fluid passing through the pump. Substituting gH_L for the last term in (4.11) and transposing it to the left hand side, we have

$$gH = g(H_{in} - H_L) = \frac{V_d^2 - V_s^2}{2} + (z_d - z_s)g + \frac{p_d - p_s}{\rho} \tag{4.12}$$

where the output head H is defined as the input head less the hydraulic loss. The right hand side of (4.12) represents the increase in the three forms of mechanical energy, viz., kinetic, potential, and flow work. Typically, only the last term need be considered in computing the output head, so that (4.12) becomes

$$H = \frac{p_d - p_s}{\rho g} \tag{4.13}$$

According to Csanady (1964) the hydraulic loss can be expressed in terms of loss coefficients k_d and k_r for the diffuser and rotor, respectively, and the corresponding kinetic energies, so that

$$gH_L = \frac{k_d V_{2'}^2}{2} + \frac{k_r W_{2'}^2}{2} \tag{4.14}$$

Equation (4.14) is used in Appendix C to determine the value of $V_{u2'}/U_2$ which gives the minimum value of hydraulic loss. For typical values of flow coefficient, i.e., 0.05–0.20, the optimum value of $V_{u2'}/U_2$ is approximately 0.5. Although (4.14) is useful in analyzing losses, H_L is usually obtained from the hydraulic efficiency, which is defined by

$$\eta_H = \frac{H}{H_{in}} = \frac{H_{in} - H_L}{H_{in}} \tag{4.15}$$

The *Pump Handbook* (1976) offers the following correlation of experimental data for hydraulic efficiency:

$$\eta_H = 1 - \frac{0.8}{Q^{1/4}}$$

(4.16)

where Q is capacity in gallons per minute. According to (4.16) the hydraulic losses, represented by $1 - \eta_H$, vary from 30 percent for pumps of 50 gpm capacity to 8 percent for pumps of 10,000 gpm capacity.

Outside the impeller, where no through flow occurs, the fluid is forced to move tangentially and radially. This circulatory motion of unpumped fluid results in an additional (disk friction) loss. A different, but equally nonproductive, use of energy occurs because of a reverse flow (leakage) from the high pressure region near the impeller tip to the low-pressure region near the inlet. The latter effect is the reason for the introduction of the volumetric efficiency η_v, defined as

$$\eta_v = \frac{\dot{m}}{\dot{m} + \dot{m}_L}$$

(4.17)

where $\dot{m}_L$ is the mass rate of leakage, and $\dot{m}$ is the mass rate of flow actually discharged from the pump.

Because of the loss of mechanical energy by the several mechanisms mentioned above, the head H, i.e., the net mechanical energy added to the fluid in the pump as determined by measurement, is less than the head computed from (4.9).

Usually, the practical performance parameter as determined by test is the overall pump efficiency η, defined by

$$\eta = \frac{\dot{m}gH}{P}$$

(4.18)

where P is the power of the motor driving the pump as determined by dynamometer test. The so-called total head H is determined from the steady-flow energy equation after experimentally evaluating the mechanical energy terms at the suction and discharge sides of the pump.

The mechanical efficiency η_m accounts for frictional losses occurring between moving mechanical parts, which are typically bearings and seals, as well as for disk friction, and is defined by

Table 4.1   Constants for Equation (4.21)

N_s	C	n
500	1.0	0.50
1000	0.35	0.38
2000	0.091	0.24
3000	0.033	0.128

$$\eta_m = \frac{(\dot{m} + \dot{m}_L)\, gH_{in}}{P} \tag{4.19}$$

Substitution of (4.9), (4.15), and (4.18) into (4.19) yields the simple relationship

$$\eta = \eta_m \eta_v \eta_H \tag{4.20}$$

The *Pump Handbook* provides data on volumetric efficiency which is correlated by

$$\eta_v = 1 - \frac{C}{Q^n} \tag{4.21}$$

where C and n are constants which depend on the dimensional specific speed N_s. Some values of these constants are presented in Table 4.1. Equation (4.21) shows that volumetric efficiencies range from 0.99 for large pumps to 0.85 for pumps of low capacity.

The mechanical efficiency can be calculated by formulating disk and bearing friction forces or from a knowledge of the other efficiencies. The overall pump efficiency can be obtained from Table 2 in Appendix A, and the other efficiencies can be calculated from equations (4.16) and (4.21). Thus, the mechanical efficiency is the only unknown in (4.20). The process is illustrated in Example Problem 4.1.

4.4   Performance Characteristics

Characteristic curves for a given pump are determined by test, and they consist primarily of a plot of head H as a function of volume rate of flow

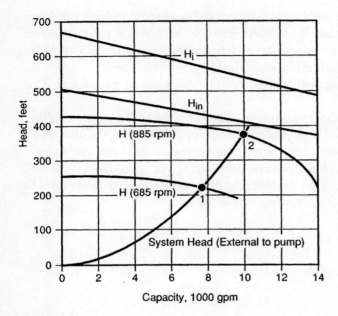

Figure 4.4 Pump characteristics.

Q. A typical characteristic curve is shown schematically in Figure 4.4. The theoretical head from (4.22) is also shown in Figure 4.4. The actual curve is displaced downward as a result of the losses of mechanical energy previously discussed. However, (4.22) provides the engineer with the upper limit of performance which can be achieved, since it does not account for losses. If the speed is increased, (4.22) indicates that the curve will shift upward, and vice versa.

Expressing (4.22) in terms of Q, we have

$$H_i = U_2 \frac{U_2 - Q \cot \beta_2 / A_2}{g} \tag{4.22}$$

Dividing (4.14) by the square of twice the tip speed ND_2, we obtain

$$\frac{gH_i}{N^2 D_2^2} = 0.5 \left(0.5 - \frac{D_2 Q \cot \beta_2}{\pi b N D_2^3} \right) \tag{4.23}$$

Equation (4.23) indicates a functional relationship between head coefficient $gH/N_2D_2^2$ and flow coefficient Q/ND_2^3, which is independent of speed. The actual performance curves, when plotted nondimensionally, also show a functional relationship which is independent of speed; i.e., data for different rotor speeds will collapse into a single-head coefficient-flow coefficient curve.

We can predict the approximate value of head or flow rate resulting from a change of speed if we assume that the operating state, i.e., the values of head and flow coefficients, are the same before and after the change. Referring to Figure 4.4 and considering a change of speed from N_1 to N_2, the operating state point on the characteristic plot moves from position 1 to position 2. Since we are assuming similar flows,

$$\frac{H_1}{N_1^2} = \frac{H_2}{N_2^2} \tag{4.24}$$

and

$$\frac{Q_1}{N_1} = \frac{Q_2}{N_2} \tag{4.25}$$

Equations (4.24) and (4.25) express the pump (or fan) laws. Manipulation of these equations yields

$$\frac{H_1}{Q_1^2} = \frac{H_2}{Q_2^2} \tag{4.26}$$

which states that H is proportional to Q^2. The latter relation is approximately that followed by the external system to which the pump is attached, assuming that no changes have been made in it. Thus the path from 1 to 2 for a simple change of speed is roughly the locus of similar states, and this fact makes the pump (or fan) laws extremely useful.

As indicated by (4.18) overall efficiency varies with flow rate, and it is required for the computation of brake power. Referring to Figure 4.5, which is a typical variation, we see that efficiency varies with flow rate from zero at no flow to a maximum value η_{max} near the highest flow rate. The actual value of η_{max} varies from 70 to 90 percent, depending primarily upon the

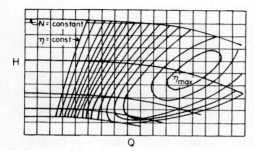

Figure 4.5 Equal-efficiency contours for centrifugal pumps.

design flow capacity. Machines handling large flows have higher maximum efficiencies, since frictional head loss decreases proportionately with large flow area. On the other hand, machines of high head and low flow, i.e., low specific speed, tend to have lower efficiencies. High head is associated with large-diameter or high-speed impellers, which increases disk-friction losses considerably, and low flow implies higher proportional head loss associated with smaller flow area. Characteristically, the latter-type machine is of the radial-flow design, while the former is classified as a mixed-flow design. Although the choice of a specific speed may be dictated by design requirements, it is worth noting that test results show that centrifugal pumps with specific speeds between 0.7 and 1.0 seem to have the highest maximum efficiencies (for example, see Church, 1972).

The head-capacity (H-Q) curves can be altered at the high-flow end by the occurrence of the phenomenon of cavitation. This process consists of the formation and collapse of vapor bubbles, which occur when the fluid pressure falls below the vapor pressure. Outward flow in the impeller passage, which is accompanied by pressure rise, results in a collapse of the bubble. Acceleration of fluid surrounding the bubble, which is required to fill the void left by the vapor, results in losses and pressure waves which cause damage to solid-boundary materials. Since the energy transfer per unit weight is reduced by the presence of vapor, the head-capacity curve falls off at the flow corresponding to the beginning of cavitation.

To avoid cavitation, the net positive suction head (NPSH), defined as the atmospheric head plus the distance of liquid level above pump centerline minus the friction head in suction piping minus the gauge vapor pressure, is maintained above a certain critical value. A critical specific speed S_c defined as

$$S_c = \frac{NQ^{1/2}}{[g(NPSH)_c]^{3/4}} \tag{4.27}$$

is used to determine the lowest safe value of NPSH. For single-suction water pumps Shepherd (1956) gives $S_c = 3$, and for double-suction pumps he gives $S_c = 4$. These form useful rules of thumb for the avoidance of cavitation by designers and users of centrifugal pumps.

The effect of high fluid viscosity on pump performance can be determined through the use of correction factors for head, capacity, and overall efficiency. The factors are ratios of head, capacity, or efficiency for viscous fluid pumping to the same parameter with water as the pumped fluid; thus,

$$c_H = \frac{H_{vis}}{H} \tag{4.28}$$

$$c_Q = \frac{Q_{vis}}{Q} \tag{4.29}$$

$$c_E = \frac{\eta_{vis}}{\eta} \tag{4.30}$$

Table 4.2 illustrates the effect of high viscosity on the head, capacity, and efficiency factors for the case of a centrifugal pump having a design

Table 4.2 Effect of Kinematic Viscosity on the Performance of a Centrifugal Pump with H = 70 ft, Q = 2400 gpm

Kinematic viscosity (centistokes)	c_E	c_Q	c_H
5	0.99	1.0	1.0
10	0.97	1.0	1.0
20	0.92	1.0	0.99
32	0.90	1.0	0.98
65	0.84	1.0	0.97
132	0.78	0.99	0.94
220	0.71	0.97	0.93

capacity of 2400 gpm at an output head of 70 feet of water. The table shows that equations and graphs developed from water-pump tests can be used to accurately predict head and capacity of pumps handling fluids with 100 times the viscosity of water. On the other hand, the efficiency decreases dramatically with increased viscosity. This effect can be explained by a significant increase in disk friction, stemming from the fact that disk friction power is proportional to the one-fifth power of kinematic viscosity.

Selected values of viscosities of liquids are presented in Table 1 of Appendix D. Values of c_H, c_Q, and c_E for an engine oil are given in Table 2 of Appendix D. Analysis of viscous pumping is facilitated through the use of charts prepared by the Hydraulic Research Institute, which are reprinted in the *Pump Handbook*.

4.5 Design of Pumps

Requirements for a pump comprise the specification of head, capacity, and speed. This section deals with the application of principles to the problem of the determination of the basic dimensions of the impeller and casing. The process outlined below would enable the engineer to carry out a preliminary design to which the detailed mechanical design could be added, or to select a suitable pump from commercially available machines.

The impeller design can be started by computing the required specific speed and using this value to determine efficiency from available test data plotted in the form of η as a function of N_s with Q as the parameter. Brake power calculated from (4.18) is then used to determine shaft torque from

$$T = \frac{P}{N} \tag{4.31}$$

The shaft torque can be used to determine the shaft diameter through the use of a formula for stress in a circular bar under torsion. As shown in Figure 4.6, the hub diameter is larger than the shaft diameter and the shaft may pass through the entire hub. For single-suction pumps, the shaft may end inside the hub and thereby not pierce the eye of the impeller.

The double-suction impeller shown in Figure 4.6 takes half of the flow in each side. The double-suction type is used to maintain low fluid speeds at the impeller eye and to avoid abrupt turning of the fluid when shroud diameters are large. The shroud diameter should not exceed half of the

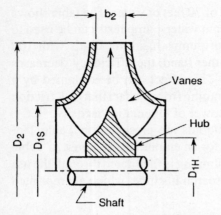

Figure 4.6 Double-suction centrifugal impeller.

impeller diameter. Each side of the double-suction impeller should be treated as a single-suction impeller when determining overall, hydraulic, or volumetric efficiency; thus, the specific speed for efficiency determination is based on Q/2, rather than Q. On the other hand, when power or tip blade width is to be determined, or when the Cordier diagram is being used to determine the specific diameter, the full flow rate Q should be used.

The impeller tip speed and the impeller diameter should be determined from the head equation,

$$H = \frac{\eta_H U_2^2 (V_{u2'}/U_2)}{g}$$

(4.32)

Using the results from Appendix C, a value of 0.5–0.55 is substituted for $V_{u2'}/U_2$ in (4.32). The hydraulic efficiency is computed from (4.16); thus, U_2 is the only unknown in (4.32) and is readily calculated for the specified head. The impeller diameter is obtained using $D_2 = 2U_2/N$.

The *Pump Handbook* recommends that the flow coefficient be chosen in the range

$$\frac{N_s}{21600} < \varphi_2 < \frac{N_s}{15900} + 0.019$$

(4.33)

where N_s is the dimensional specific speed. The selected value of flow coefficient is used to determine the meridional velocity using the defining relation,

$$W_{m2} = \varphi_2 U_2 \tag{4.34}$$

Using values of constants from Table 4.1 the volumetric efficiency is determined from (4.21). The impeller flow rate $Q + Q_L$ is determined from (4.17) by dividing $\dot{m}$ by ρ to obtain Q. The impeller flow rate is used to determine the vane tip width from

$$b_2 = \frac{Q + Q_L}{\pi D_2 W_{m2}} \tag{4.35}$$

The *Pump Handbook* recommends the following equations for the calculation of the shroud diameter:

$$k = 1 - \left(\frac{D_{1H}}{D_{1S}}\right)^2 \tag{4.36}$$

and

$$D_{1S} = 4.54 \left(\frac{Q + Q_L}{kN \tan \beta_{1s}}\right)^{1/3} \tag{4.37}$$

where the shroud diameter is given in inches when N is in rpm and Q is in gpm. Equation (4.37) applies to single-suction impellers. If it is to be used for double-suction impellers, then $(Q + Q_L)/2$ should be used.

The hub-tip ratio used in (4.36) is selected by the designer and can have values from zero to more than 0.5. The constant k approaches unity as the hub-tip ratio decreases to zero and can be taken as unity to approximate the shroud diameter. The hub diameter must exceed the shaft diameter if one passes through the eye. In such cases, the ratio may be taken as 0.5. An optimum hub-tip ratio for minimizing the relative velocity in the eye can be obtained by differentiation of W_{1S} with respect to D_{1S}.

The inlet vane angle at the shroud is selected by the designer from the range of values recommended by the *Pump Handbook*, viz., 10–25 degrees. When all values are substituted into (4.37), the shroud diameter is determined. The hub-tip ratio is then used to determine the design hub diameter.

The next phase of the preliminary design involves iteration on the vane angle β_2. The designer selects a vane angle In the range of 17–25 degrees, as is recommended by the *Pump Handbook*. Using the selected angles along with the calculated diameters, the optimum number of vanes is calculated from an equation recommended by Pfleiderer (1949) and Church (1972), viz.,

$$n_B = 6.5 \left(\frac{D_2 + D_{1S}}{D_2 - D_{1S}} \right) \sin \left(\frac{\beta_{1S} + \beta_2}{2} \right)$$

(4.38)

The optimum number of vanes for pumps lies in the range of 5 to 12.

The slip coefficient is combined with its basic definition to yield

$$\mu_s = \frac{V_{u2'}/U_2}{V_{u2}/U_2} = 1 - \frac{\pi \sin \beta_2}{n_B} \left(\frac{U_2}{V_{u2}} \right)$$

(4.39)

This equation can be solved for V_{u2}/U_2. The assumed vane angle can now be tested by calculating a new vane angle from

$$\beta_2 = \tan^{-1} \left| \frac{W_{m2}/U_2}{1 - V_{u2}/U_2} \right|$$

(4.40)

When the new value of vane angle agrees with the assumed value, the design is complete, in that the basic impeller dimensions will have been determined.

The fluid exits the impeller with tangential and radial components of absolute velocity and is collected and conducted to the discharge of the pump by the volute or scroll portion of the casing (Figure 4.7). The volute is usually in the form of a channel of increasing cross-sectional area. The volute begins at the tongue with no cross-sectional area and ends at the discharge nozzle. At any angle φ, measured from the tongue, the flow rate is $(\varphi/360)Q$. The angular momentum of the exit flow, $V_{u2'}r_2$, is conserved, so the distribution is approximately

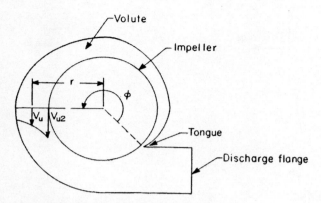

Figure 4.7 Volute of a pump.

$$V_u r = \text{constant} \tag{4.41}$$

The angle φ corresponding to each radial coordinate r_3 is determined from the integrated volume flow equation

$$\frac{\varphi Q}{360} = w V_{u2'} \, r_2 \ln \frac{r_3}{r_2} \tag{4.42}$$

If the channel width w is variable, as in a channel of circular cross section, then the governing relation should be

$$\frac{\varphi Q}{360} = V_{u2'} \, r_2 \int_{r_2}^{r_3} \frac{w}{r} \, dr \tag{4.43}$$

The so-called discharge nozzle, which is really a diffuser, joins the volute to the discharge flange of the pump. For water the nozzle is typically sized to produce a discharge velocity of 25 ft/s. A radial diffuser may be added between the impeller and the volute for high-pressure pumps. This may take the form of a space of constant width without vanes, or it may include vanes forming diverging passages aligned with the absolute velocity vector.

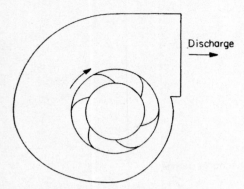

Figure 4.8 Centrifugal fan.

4.6   Fans

Fans produce very small pressure heads measured in inches of water pressure differential, and of course are employed to move air or other gases. A compressor also handles gases, but with large enough pressure rises that significant fluid density changes occur; i.e., if density is increased by 5 percent, then the turbomachine may be called a compressor.

A centrifugal fan, as compared with a pump, requires a much smaller increase in impeller blade speed, i.e., a smaller radius ratio R_2/R_1, as may be inferred from (4.7). It requires a volute, of course, but no diffuser is needed to enhance pressure rise. The flow passages between impeller vanes are quite short, as indicated in Figure 4.8.

The analysis and design of the impeller proceeds as with the centrifugal pump. The small changes of gas density are ignored, and the incompressible equations are applied as with pumps. Performance curves are qualitatively the same as for pumps, except that the units of head are customarily given in inches of water, and those of capacity are typically in cubic feet per minute.

Other differences are that both total head and static (pressure) head are usually shown on performance curves, and a fan static efficiency, based on (4.18), is calculated using static head $(P_2 - P_1)/pg$ in place of total head H. Similarity laws for pumps are applied and are known as fan laws; these are represented by (4.24) and (4.25).

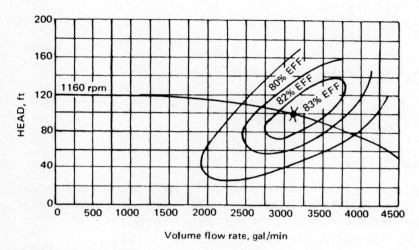

Figure 4.9 Head-capacity curve for a pump.

4.7 Examples

Example Problem 4.1

Determine the overall, hydraulic, volumetric, and mechanical efficiency for a centrifugal pump having a capacity of 1000 gpm and a dimensional specific speed of 1600.

Solution: Obtain the overall pump efficiency from Table 2 of Appendix A.

$$\eta = 0.83$$

Calculate the hydraulic efficiency using equation (4.16).

$$\eta_H = 1 - \frac{0.8}{Q^{\frac{1}{4}}} = 1 - \frac{0.8}{(1000)^{\frac{1}{4}}} = 0.858$$

Determine the constants for equation (4.21) by interpolating between values in Table 4.1. $C = 0.195$; $n = 0.296$. Calculate volumetric efficiency from (4.21).

$$\eta_v = 1 - \frac{0.195}{(1000)^{0.296}} = 0.975$$

Finally, determine the mechanical efficiency from (4.20).

$$\eta_m = \frac{\eta}{\eta_H \eta_v} = \frac{0.83}{(0.858)(0.975)} = 0.99$$

Example Problem 4.2

A single-suction centrifugal pump runs at 885 rpm while delivering water at the rate of 10,000 gpm. Determine the ideal, input, and output heads, if the impeller diameter is 38 inches, and the tip vane angle is 21.6°.
Solution: Convert the rotational speed from rpm to rad/s.

$$N = (885)(\pi/30) = 92.68 \text{ rad/s}$$

Determine the tip speed.

$$U_2 = ND_2/2 = (92.68)(38)/(24) = 146.7 \text{ fps}$$

Find W_{m2}. Since the impeller tip width is not known, choose a value for φ_2 in the range of design values prescribed in Section 4.5. The head is unknown but will be determined; hence, an assumed value of dimensional specific speed can be checked at the conclusion. Assume $N_s = 1000$. Use the upper limit for the flow coefficient, viz.,

$$\varphi_2 = 1,000/15,900 + 0.019 = 0.082$$

and determine the meridional component.

$$W_{m2} = (146.7)(0.082) = 12 \text{ fps}$$

Calculate V_{u2}. Refer to Figures 4.2 and B-2 (in Appendix B).

$$V_{u2} = U_2 - W_{m2} \cot \beta_2 = 146.7 - (12) \cot 21.6° = 116.4 \text{ fps}$$

Calculate the ideal head using equation (4.1).

$$H_i = U_2 V_{u2}/g = (146.7)(116.4)/32.174 = 531 \text{ ft}$$

Calculate the slip coefficient. From Section 4.5 we learn that the optimum number of vanes for a centrifugal pump is between 5 and 12. A conservative choice would be 6 vanes. Let $n_B = 6$. Now apply equation (4.8).

$$\mu_s = 1 - \frac{(3.14159)(146.7) \sin 21.6°}{(116.4)(6)} = 0.757$$

Calculate $V_{u2'}$, using the definition of slip coefficient.

$$V_{u2'} = \mu_s V_{u2} = (0.757)(116.4) = 88.1 \text{ fps}$$

Calculate the input head using equation (4.9).

$$H_{in} = (146.7)(88.1)/32.174 = 401.7 \text{ ft}$$

Calculate the hydraulic efficiency using equation (4.16).

$$\eta_H = 1 - \frac{0.8}{(10,000)^{1/4}} = 0.92$$

Calculate the output head using equation (4.15).

$$H = (401.7)(0.92) = 369.6 \text{ ft}$$

Check assumed specific speed.

$$N_s = \frac{NQ^{1/2}}{H^{3/4}} = \frac{(885)(10,000)^{1/2}}{(369.6)^{3/4}} = 1050$$

The assumed value is acceptable.

Example Problem 4.3

The single-suction centrifugal pump whose characteristics are shown in Figure 4.9 is operating at the design point, i.e., $Q = 3100$ gpm, $H = 100$ ft and $N = 1160$ rpm. The suction pipe connecting the pump suction to the supply reservoir has a diameter D_{su} of 8 inches and a length L_{su} of 10 feet. The pipe lifts water from a reservoir 10 feet below the centerline of the pump. The free surface of the reservoir is at 14.7 psia. Determine the suction specific speed, and assess the adequacy of the design.

Solution: The net positive suction head is given by

$$\text{NPSH} = \frac{P_{atm}}{\rho g} + z_R - h_f - \frac{P_{vap}}{\rho g}$$

We are given that $z_R = -10$ ft. The specific weight of water ρg is 62.4 lb/ft^3. The vapor pressure of water is obtained from a table of thermodynamic properties, e.g., from Moran and Shapiro (1988), which gives $P_{vap} = 0.5073$ psi at 80°F. The head loss h_f is determined from

$$h_{f.} = \frac{fL_{su}V_{su}^2}{2gD_{su}}$$

The average velocity V_{su} in feet per second in the suction pipe is

$$V_{su} = \frac{4Q}{\pi D_{su}^2} = \frac{4(3100/448.8)}{\pi (8/12)^2} = 19.79 \text{ fps}$$

The kinematic viscosity of water may be taken as 1.0 cs (centistoke), which is easily converted to English units using the conversion factor from Table 1 of Appendix A.

$$\nu = 1/92,903 = 0.000010764 \text{ ft}^2/s$$

The Reynolds number of the pipe flow is

$$\text{Re} = \frac{V_{su}D_{su}}{\nu} = \frac{19.79(8/12)}{0.000010764} = 1,225,565$$

The friction factor f is taken from the Moody diagram, e.g., from Fox and McDonald (1985) and is f = 0.0112. The head loss can now be calculated as

$$h_f = \frac{(0.0112)(10)(19.79)^2}{2(32.174)(8/12)} = 1.02 \text{ ft}$$

Substituting into the equation for NPSH, we have

$$NPSH = \frac{14.7(144)}{62.4} + (-10) - 1.02 - \frac{0.5073(144)}{62.4} = 19.56 \text{ ft}$$

The definition of the suction specific speed S is

$$S = \frac{NQ^{\frac{1}{2}}}{[g(NPSH)]^{\frac{3}{4}}}$$

We calculate S stepwise and obtain

$$N = \frac{1160}{9.54929} = 121.47 \text{ rad/s}$$

$$Q = \frac{3100}{448.8} = 6.9073 \text{ cfs}$$

$$g(NPSH) = 32.174(19.56) = 629.3 \text{ ft}^2/\text{s}^2$$

$$S = \frac{121.47(6.9073)^{\frac{1}{2}}}{(629.3)^{\frac{3}{4}}} = 2.54$$

Since the suction specific speed S is less that its critical value S_c, we can assume that cavitation will not occur and that the design, i.e., the elevation of the pump above the supply reservoir, is acceptable.

Example Problem 4.4

Estimate the performance of the centrifugal pump in Example Problem 4.3 for viscous pumping of a fluid with $v = 150$ centistokes at the same speed.
Solution: Pump performance with water is given as

$$Q = 3100 \text{ gpm}$$
$$H = 100 \text{ ft}$$
$$\eta = 0.83$$

The correction factors in Table 2 in Appendix D are for a viscosity of 176 cs and will give somewhat more conservative predictions of performance with $v = 150$ cs; however, we will use them without adjustment. Interpolating in Table 2 in Appendix D, we compute

$$c_E = 0.72$$
$$c_Q = 0.97$$
$$c_H = 0.94$$

Predictions of performance based on these factors are the following:

$$\eta_{vis} = c_E\eta = 0.72(0.83) = 0.60$$

$$Q_{vis} = c_Q Q = 0.97(3100) = 3007 \text{ gpm}$$

$$H_{vis} = c_H H = 0.94(100) = 94 \text{ ft}$$

Example Problem 4.5

The specifications of a double-suction centrifugal water pump are the following:

$$Q = 2400 \text{ gpm}$$
$$H = 70 \text{ ft}$$
$$N = 870 \text{ rpm}$$

Find: D_2, b_2, D_{1S}, D_{1H}, β_2, β_{1S}, and n_B for the impeller.
Solution: Choose $V_{u2'}/U_2 = 0.5$ on the basis of the results of Appendix C. Calculate hydraulic efficiency from (4.16).

$$\eta_H = 1 - \frac{0.8}{(1200)^{1/4}} = 0.864$$

Solve the head equation (4.32) for U_2.

$$U_2 = \left(\frac{32.174(70)}{0.5(0.864)}\right)^{1/2} = 72.2 \text{ fps}$$

Determine D_2.

$$D_2 = \frac{2U_2}{N} = \frac{2(72.2)}{(870/9.549)} = 1.585 \text{ ft} = 19.02 \text{ inches}$$

Round off calculated diameter. Choose $D_2 = 19$ inches.
Recalculate U_2.

$$U_2 = \frac{870(19)}{9.54(24)} = 72.17 \text{ fps}$$

Calculate dimensional specific speed. Use Q/2.

$$N_s = \frac{870(1200)^{\frac{1}{2}}}{(70)^{\frac{3}{4}}} = 1245$$

Determine flow coefficient from (4.33). Upper limit is

$$\varphi_2 = \frac{1245}{15900} + 0.019 = 0.0973$$

Choose $\varphi_2 = 0.09$
Determine constants from Table 4.1 by interpolation.

$$C = 0.287 \quad n = 0.346$$

Calculate volumetric efficiency from (4.21).

$$\eta_v = 1 - \frac{0.287}{(1200)^{0.346}} = 0.975$$

Calculate impeller flow rate using (4.17) and dividing by fluid density.

$$Q + Q_L = \frac{Q}{\eta_v} = \frac{2400}{0.975} = 2461 \text{gpm}$$

Calculate meridional velocity at tip using (4.34).

$$W_{m2} = \varphi_2 U_2 = 0.09(72.17) = 6.495 \text{ fps}$$

Calculate tip width of vane from (4.35).

$$b_2 = \frac{Q + Q_L}{\pi D_2 W_{m2}} = \frac{(2461/448.8)}{\pi(19/12)(6.495)} = 0.1697 \text{ ft} = 2.036 \text{ in}$$

Choose $b_2 = 2.0$ inches
Choose hub-shroud ratio to be 0.5. Calculate k from (4.36).

$$k = 1 - (0.5)^2 = 0.75$$

Choose $\beta_{1S} = 17°$.
Calculate the shroud diameter using (4.37).

$$D_{1S} = 4.5 \left[\frac{(Q + Q_L)/2}{kN \tan \beta_{1S}} \right]^{1/3} = 4.5 \left[\frac{1230}{0.75(870) \tan 17°} \right]^{1/3} = 8.25 \text{ in}$$

Calculate the hub diameter using the hub-shroud ratio.

$$D_{1H} = 0.5(8.25) = 4.125 \text{ in}$$

Begin iteration by trying a value of 20° for β_2.
Calculate the number of vanes using (4.38).

$$n_B = 6.5 \frac{(19 + 8.25)}{(19 - 8.25)} \sin \left(\frac{17 + 20}{2} \right) = 5.2$$

Choose $n_B = 5$
Calculate V_{u2}/U_2 using (4.8).

$$\mu_s = \frac{V_{u2'}/U_2}{V_{u2}/U_2} = 1 - \frac{\pi \sin \beta_2}{n_B V_{u2}/U_2}$$

Solve the above equation for V_{u2}/U_2 and adjust $V_{u2'}/U_2$ from 0.5 to 0.54.

$$\frac{V_{u2}}{U_2} = 0.54 + \frac{\pi \sin 20°}{5} = 0.755$$

Check assumed value of β_2 using (4.40).

$$\beta_2 = \tan^{-1} \frac{0.09}{0.245} = 20.17°$$

Agreement satisfactory. Choose $\beta_2 = 20°$.
Calculate μ_s using (4.8). $\mu_s = 0.715$.
Check head produced.

$$H = \frac{\eta_H \, \mu_s U_2^2 (V_{u2}/U_2)}{g}$$

$$H = \frac{0.864(0.715)(72.17)^2(0.755)}{32.174} = 75.5 \text{ ft}$$

The result is a little higher than specified, but the design is acceptable.

References

Church, A. H. 1972. *Centrifugal Pumps and Blowers*. Krieger, Huntington, New York.

Csanady, G. T. 1964. *Theory of Turbomachines*. McGraw-Hill, New York.

Fox, R. W. and A. T. McDonald. 1985. *Introduction to Fluid Mechanics*. John Wiley and Sons, New York.

Karassik, I. J., et al. 1976. *Pump Handbook*. McGraw-Hill, New York.

Moran, M. J., and H. N. Shapiro. 1988. *Fundamentals of Engineering Thermodynamics*. John Wiley & Sons, New York.

Pfleiderer, C. 1949. *Die Kreiselpumpen*. Springer-Verlag, Berlin.

Shepherd, D. G. 1956. *Principles of Turbomachinery*. MacMillan, New York.

Problems

4.1. A centrifugal water pump has the characteristic curves shown in Figure 4.9. Using the 1160-rpm curve plot the characteristics (H versus Q) for a geometrically similar pump having twice the speed and half the diameter (of the rotor). Show calculations that were used to obtain the coordinates plotted.

4.2. Calculate the power required to drive the original pump at 1160 rpm at a flow rate of 3100 gal/min. Also determine the specific speed (unitless).

4.3. For a pump impeller with a diameter D_2 of 1.326 ft and axial width $b_2 = 2$ in. Determine the velocity diagram at the exit of the rotor for the conditions in Problem 4.2. Vane angle $\beta_2 = 25°$. There are seven vanes.

4.4. Determine the principal dimensions of a centrifugal pump which can deliver 3100 gal/min of water at a 100 ft head. The speed is 1160 rpm.

4.5. A centrifugal kerosene pump with backward-curved vanes runs at 1876 rpm and delivers 0.678 cu ft/s. The tip width of the vanes is $b_2 = 0.498$ in. The tip diameter $D_2 = 0.505$ ft. The head $H = 28$ feet. The density of the kerosene is 50.8 lb/cu ft.
  a) draw and label the velocity diagram
  b) solve for U_2 in fps
  c) find $V_{u2'}$, in fps
  d) find V_{m2} in fps
  e) find $\beta_{2'}$ in degrees

4.6. Derive equation (4.2), and note that it applies only in the case of solid body rotation of the fluid, i.e., without flow through the pump.

4.7. Derive equation (4.5) using the outlet velocity triangle in Figure 4.2.

4.8. Show that $H_i = (U_2^2/g)$ applies to centrifugal pumps which have an infinite number of vanes but have neither throughflow $(Q = 0)$ nor friction loss by
  a) using equation (4.22)
  b) using equations (2.34) and (4.7)
  Hint: As $Q \to 0$, $W_1 \to -U_1$ and $V_2 \to U_2$. In equation (2.34) substitute $H = H_i$.

4.9. A centrifugal water pump produces a head of 70 feet while delivering 2400 gpm at a speed of 870 rpm. The impeller tip diameter is 19 inches, and the tip vane width is 2 inches. The pump is of the double-suction type with the shroud diameter at impeller eye of 9.50 inches penetrated by a shaft having a diameter of 2.20 inches with a hub diameter of 4.75 inches. The impeller has six backward-curved vanes. Determine:
  a) the specific speed assuming 1200 gpm entering each side of the impeller
  b) the overall pump efficiency
  c) the hydraulic efficiency
  d) the volumetric efficiency
  e) the mechanical efficiency
  f) the required shaft power for operation at the design speed
  g) the required shaft power for operation at 1200 rpm

4.10. Construct the actual and ideal velocity triangles for the pump described in Problem 4.9. Find:

    a) the axial component of velocity entering the eye of the impeller
    b) the flow coefficient at the impeller exit
    c) $V_{u2'}/U_2$ for the actual diagram
    d) the fluid angle at the impeller tip $\beta_{2'}$
    e) V_{u2}/U_2 for the ideal diagram
    f) the slip coefficient
    g) the vane angle at the impeller tip β_2

4.11. A centrifugal water pump is to be designed to create a pressure rise of 30 psi while delivering 1000 gpm at a speed of 2000 rpm. Using a flow coefficient of 0.122, a tip vane angle of 25°, and six impeller vanes, find:

    a) the expected overall efficiency
    b) the expected hydraulic efficiency
    c) the expected slip coefficient
    d) the appropriate impeller tip diameter
    e) the axial tip width of the impeller vane

4.12. An impeller of a centrifugal water pump used in irrigation of farm lands has the following dimensions:

    Tip vane angle, $\beta_2 = 25°$
    leading-edge vane angle, $\beta_1 = 12°$
    Shroud diameter at eye = 4.875 inches
    Vane leading-edge diameter = 4.875 inches
    Impeller tip diameter = 8.625 inches
    Vane axial width at tip = 1.375 inches
    Vane axial width at leading edge = 1.50 inches
    Number of cylindrical, backward-curved vanes = 6

Assuming a design speed of 1750 rpm, find:

    a) the design capacity in gpm
    b) the design head in feet of water
    c) the drive motor power in hp

4.13. For a centrifugal pump with backward-curved vanes it may be assumed that the loss of input head H_L is given by the expression,

$$gH_L = K_d V_{2'}^2/2 + K_r(W_{2'})^2/2$$

where K_d is the loss coefficient for the diffuser, K_r is the loss coefficient for the rotor (impeller), and $V_{2'}$ and $W_{2'}$ are the actual fluid velocities. Assuming that $K_r/K_d = \frac{1}{3}$, as suggested by Csanady, and that the optimum actual velocity diagram corresponds to the minimum value of H_L/H_{in}, determine:

    a) the optimal values of $V_{u2'}/U_2$ for tip flow coefficients of 0.05, 0.1, and 0.2

    b) the corresponding optimal fluid angles at the impeller tip

4.14. Consider a centrifugal pump designed for operation at a specific speed in the 1000–3000 range. Appropriate midrange values of flow coefficient and vane tip angle are $\varphi_2 = 0.1$ and $\beta_2 = 22°$. For an impeller having six vanes, determine:

    a) V_{u2}/U_2 for the ideal tip velocity diagram

    b) the corresponding slip coefficient

    c) $V_{u2'}/U_2$ for the actual velocity diagram

    d) compare these practical values with those resulting from the analysis of Problem 4.13.

4.15. A centrifugal pump having nine backward-curved vanes and a tip diameter of 6.06 inches operates at 1500 rpm and delivers 0.6684 cfs of water. The vanes make an angle of 30 degrees with the tangential direction at the tip of the impeller. The axial width of the vane at the impeller tip is 0.5 inch. Calculate:

    a) the tip blade speed in fps

    b) V_{m2} in fps

    c) V_{u2} in fps

    d) the slip coefficient

    e) the input head in ft of water.

    f) the output head in ft water

4.16. A single-sided centrifugal gasoline pump with six backward-curved vanes runs at 1876 rpm and delivers 0.678 cu ft/s. The tip width of the vanes is $b_2 = 0.498$ inch. The vanes are cylindrical* and the shaft and hub do not extend into the eye of the impeller. The tip diameter $D_2 = 0.505$ ft, and the shroud diameter is 0.25 ft. The output head $H = 28$ ft. The density of the gasoline is 46 lb/cu ft, its kinematic viscosity is 0.000006 ft²/s and its vapor pressure is 10 psi.

    a) draw and label the velocity diagram for perfect guidance by the vanes; superimpose the diagram for a finite number of vanes

    b) find the flow coefficient

---

* The term "cylindrical" means the vanes do not extend into the impeller eye.

  c) calculate specific speed in gpm-rpm-ft units; what range of values of flow coefficient are possible for good design?
  d) find the tip vane angle in degrees
  e) calculate the dimensionless and unitless specific diameter
  f) find the input head in feet of gasoline
  g) find the ideal head in feet of gasoline
  h) find the horsepower required to drive the pump
  i) find the shaft torque at steady speed
  j) find the lowest acceptable suction pressure in psi

4.17. Determine the height that water at 80°F can be lifted by the centrifugal pump in Problem 4.4, if the pump is connected by means of an 8-inch diameter suction pipe to a supply reservoir whose free surface is maintained at a pressure of 1 atmosphere.

4.18. A three-stage, centrifugal pump operating at 38,000 rpm handles 125 lb/s of liquid hydrogen (vapor pressure = 14.7 psia; density = 4.43 lb/cu ft) at a suction pressure of 100 psia and a discharge pressure of 6300 psia. Assume that the shroud-to-tip diameter ratio is 0.5 and the hub-to-shroud diameter ratio is 0.44. Find:
  a) the impeller diameter
  b) the shroud diameter
  c) the hub diameter
  d) the mean velocity at the impeller eye
  e) the suction specific speed.

Hint: Assume equal pressure rises for the three stages.

4.19. Determine the impeller diameter and the suction specific speed of a small centrifugal cryogenic pump with the following performance requirements:

  $Q = 0.016 \, m_3/s$
  $H = 2230 \, m$
  $N = 45,000 \, rpm$
  $NPSH = 20 \, m$.

Note: The result shows that $S > S_c$; thus, an axial-flow stage ahead of the eye of the centrifugal pump is needed. This stage is called an inducer and is frequently required for high speed pumps.

4.20. Oil having a specific gravity of 0.886 and a temperature of 130°F is pumped by three centrifugal pumps arranged in parallel at the rate of 2,116,000 barrels (42 gallons = 1 barrel) per day from a suction pressure of 26 psig to a discharge pressure of 1150 psig. The pump is driven by a gas turbine at a speed of 3400 rpm. Design one impeller of a two-stage, double suction centrifugal pump for the service indicated above. Assume that the two impellers are identical.

4.21. Performance curves for a double-suction centrifugal pump indicate the following data at maximum efficiency (90 percent): H = 345 ft; Q = 20,000 gpm; N = 885 rpm. Determine the principal dimensions for such a pump.

4.22. Assume that the velocity of water entering one side of the pump in Problem 4.21 is V_{su} = 19.1 fps and that the water temperature is 80°F. Assume that $S = S_c$ and that h_f is negligible. Find the diameter of the suction pipe and the pressure of the water entering the pump at the suction flange.

Symbols for Chapter 4

A_2	flow area at impeller exit
b_2	width of vane at $r = r_2$ in pump impeller
C	constant in Equation (4.21)
c_E	efficiency correction factor for viscosity
c_H	head correction factor for viscosity
c_Q	capacity correction factor for viscosity
D_{1S}	shroud diameter
D_{1H}	hub diameter
D_2	impeller tip diameter
D_s	specific diameter
D_{su}	diameter of suction pipe
e_d	specific internal energy of fluid at discharge flange of pump
e_s	specific internal energy of fluid at suction flange of pump
f	friction factor
g	gravitational acceleration
H	output head
H_{vis}	output head with viscous pumping
H_i	ideal head
H_{in}	input head
H_L	hydraulic loss in impeller and diffuser
h_d	specific enthalpy of fluid at discharge flange of pump
h_s	specific enthalpy of fluid at suction flange of pump
h_f	friction head loss in suction pipe
k	function of hub-tip ratio

k_d	loss coefficient in diffuser of pump
k_r	loss coefficient in impeller (or rotor) of pump
L_{su}	length of suction pipe
$\dot{m}$	mass flow rate of fluid at discharge flange of pump
$\dot{m}_L$	mass flow rate of fluid leaked around the outside of the impeller from the high to the low pressure regions
N	rotor speed
NPSH	net positive suction head
N_s	specific speed
n_B	number of vanes (or blades) in the impeller
n	constant in equation (4.21)
P	power to impeller shaft
P_{atm}	atmospheric pressure
p_d	static pressure of fluid at discharge flange of pump
p_s	static pressure of fluid at suction flange of pump
p_{vap}	vapor pressure of fluid
Q	capacity, i.e., flow rate delivered by pump at discharge flange
Q_L	volume flow rate of leaked fluid
Q_{vis}	flow rate with viscous pumping
Re	Reynolds number
S	suction specific speed
S_c	critical suction specific speed
U_1	impeller speed at vane leading edge
U_2	impeller speed at vane tip
V_d	average velocity at discharge flange of pump
V_s	average velocity at suction flange of pump
V_{su}	average velocity in suction pipe
V_1	absolute velocity at leading edge of impeller vane
V_{u1}	tangential component of V_1
V_{m1}	meridional component of V_1
V_2	absolute velocity leaving the vane of an impeller with an infinite number of vanes
V_2'	absolute velocity leaving the vane of an impeller with a finite number of vanes
V_{m2}	meridional component of V_2 or $V_2' = W_{m2}$
V_{u2}	tangential component of V_2
W_1	relative velocity at leading edge of impeller vane
W_2	relative velocity leaving the vane of an impeller with an infinite number of vanes

$W_{2'}$ relative velocity leaving the vane of an impeller with a finite number of vanes

W_{m2} meridional component of W_2, W_2', V_2, or V_2'

W_{1s} relative velocity at shroud

z_d elevation of the center of the discharge flange above (positive) or below (negative) the centerline of the shaft of the pump

z_R elevation of the free surface of the supply reservoir above (positive) or below (negative) the centerline of the pump

z_s elevation of the center of the suction flange above (positive) or below (negative) the centerline of the shaft of the pump

β_1 angle between W_1 and U_1

β_{1S} angle between W_{1S} and U_{1S}

β_2 angle between W_2 and U_2; also the vane angle at trailing edge of vane

$\beta_{2'}$ angle between W_2' and U_2; actual fluid angle

η overall pump efficiency

η_H hydraulic efficiency

η_m mechanical efficiency

η_v volumetric efficiency

η_{vis} overall efficiency with viscous pumping

μ_s slip coefficient

ρ fluid density

φ flow coefficient

φ volute angle

φ_2 flow coefficient at impeller exit = W_{m2}/U_2

5 Centrifugal Compressors

5.1 Introduction

Although centrifugal compressors are slightly less efficient than axial-flow compressors, they are easier to manufacture and are thus sometimes preferred. Additionally, a single stage of a centrifugal compressor can produce a pressure ratio of 5 times that of a single stage of an axial-flow compressor. Thus, the centrifugal machine finds application in ground-vehicle power plants, auxiliary power units, and other small units.

The parts of a centrifugal compressor are the same as those of a pump, namely, the impeller, the diffuser, and the volute. The basic equations developed in Chapter 4 apply to compressors with the difference that density does increase, and we must consider the thermodynamic equation of state of a perfect gas in the detailed calculations. The main difference in carrying out a compressor analysis, as opposed to a pump analysis, is the appearance of an enthalpy term in place of the flowwork or pressure-head term. It is convenient to use both total and static enthalpy, denoted by h_o and h, respectively. Thus, energy transfer E is given by

$$E = \eta_m(h_{o3} - h_{o1}) \tag{5.1}$$

as well as by

$$E = U_2 V_{u2'} \tag{5.2}$$

where η_m denotes the mechanical efficiency.

Equations like (5.1) and (5.2) used in the same analysis require care in handling units, since the enthalpy difference in (5.1) may carry units such as Btu/lb, whereas (5.2) would carry units of velocity squared. Suitable conversion factors do not appear in the equations but must be applied in computations with them.

Since thermodynamic calculations are involved in compressor analysis and design, the h-s diagram, such as that shown in Figure 5.1, becomes useful. The state at the impeller inlet is indicated by point 1, and that at the impeller outlet by point 2. The diffuser process is indicated between points 2 and 3. The corresponding stagnation properties 01, 02, and 03 are also indicated in Figure 5.1, since kinetic energies are usually considerable. The expression for compressor efficiency appears to be somewhat different from that for pump efficiency, but, in reality, the principle of the definition is the same. Both definitions employ the ratio of the useful increase of fluid energy divided by the actual energy input to the fluid. For the compression of a gas, the useful energy input is the work of an ideal, or isentropic, compression to the actual final pressure P_3. This is calculated from

$$E_i = C_p T_{o1} \left[\left(\frac{P_{o3}}{P_{o1}} \right)^{(\gamma-1)/\gamma} - 1 \right]$$

$$(5.3)$$

which evaluates the work of the isentropic process from state 01 to state i in Figure 5.1. The compressor efficiency can be reduced to

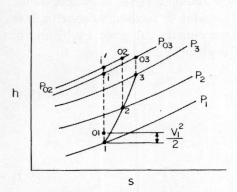

Figure 5.1 Enthalpy-entropy diagram.

$$\eta_c = \frac{T_i - T_{o1}}{T_{o3} - T_{o1}} \tag{5.4}$$

which is the ratio of E_i to E. An underlying assumption in the development of equations (5.3) and (5.4) is that there is no external work associated with the diffuser flow, nor is there any heat transfer; thus, $h_{o2} = h_{o3}$ and $T_{o2} = T_{o3}$.

The compressor efficiency, an experimentally determined quantity, is useful in predicting pressure ratios in new designs. Using (5.2), (5.3), and (5.4) we can obtain the overall pressure ratio:

$$\frac{P_{o3}}{P_{o1}} = \left(1 + \frac{U_2 V_{u2'} \eta_c}{c_p T_{o1} \eta_m}\right)^{\gamma/(\gamma - 1)} \tag{5.5}$$

Since relative eddies are present between the vanes, as in the case of centrifugal pumps, slip exists in the compressor impeller; consequently, the slip coefficient is used to calculate the actual tangential velocity component, which is given by

$$V_{u2'} = \mu_s V_{u2} \tag{5.6}$$

For compressors, however, the Stanitz equation,

$$\mu_s = 1 - \frac{0.63\pi}{n_B} \left(\frac{1}{1 - \varphi_2 \cot \beta_2}\right) \tag{5.7}$$

is used in place of equation (4.8) to calculate the slip coefficient. Several such equations are available, as is indicated in Appendix E, but the Stanitz equation is an accurate predlctor of slip coefficient for the usual range of vane angles encountered in practice, viz., $45° < \beta_2 < 90°$. Thus, the total pressure ratio for a compressor stage can be determined from a knowledge of the ideal velocity triangle at the impeller exit, the number of vanes, the inlet total temperature, and the stage and mechanical efficiencies. The mechanical efficiency is defined by equation (4.19) and accounts for frictional losses associated with bearing, seal, and disk fricton. It is assumed that the mechanical energy lost through frictional processes reappears as

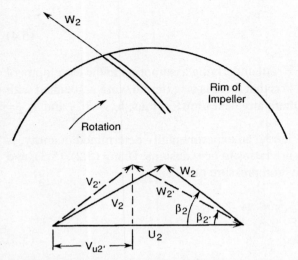

Figure 5.2 Velocity diagram at impeller exit.

enthalpy in the outflowing gas; hence, the specific shaft work into the compressor is given by E/η_m, as is shown in equation (5.1).

5.2 Impeller Design

The impeller is usually designed with a number of unshrouded blades, given by the Pfleiderer equation (4.38), to receive the axially directed fluid ($V_1 = V_{m1}$) and deliver the fluid with a large tangential velocity component $V_{u2'}$, which is less than the tip speed U_2, but it has the same sense or direction. The vanes are usually curved near the rim of the impeller, so that $\beta_2 < 90°$, but they are usually bent near the leading edge to conform to the direction of the relative velocity W_1 at the inlet.

The angle β_1 varies over the leading edge, since V_1 remains constant while U_1 (and r) varies. At the shroud diameter D_{1S} of the impeller inlet, the relative velocity W_{1S} and the corresponding relative Mach number M_{R1S} are highest. This is because the vane speed U_1 increases from hub to tip at the inlet plane, and the incoming absolute velocity V_1 is assumed to be uniform over the annulus. Referring to Figure 5.3, it is clear that $W_1 = (V_1^2 + U_1^2)^{1/2}$ and that the maximum value of W_1 occurs at the shroud diameter. It is easily shown for a fixed set of inlet operating

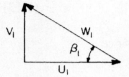

Figure 5.3 Velocity diagram at impeller inlet.

conditions, i.e., N, $\dot{m}$, P_{o1}, and T_{o1}, that the relative Mach number has its minimum where β_{1S} is approximately 32° (see Shepherd, 1956).

Referring to Figure 5.3, it is seen that a choice of relative Mach number at the shroud of the inlet vane allows the inlet design to proceed in the following manner. The acoustic speed a_1 is calculated from the inlet temperature. Next, W_1 at the shroud is calculated from

$$W_{1S} = M_{R1S}a_1 \tag{5.8}$$

where the acoustic speed is calculated from

$$a_1 = (\gamma R T_1)^{\frac{1}{2}} \tag{5.9}$$

The static temperature is given by

$$T_1 = \frac{T_{o1}}{1 + (\gamma - 1)M_1^2 / 2} \tag{5.10}$$

where the absolute inlet Mach number is defined by

$$M_1 = \frac{V_1}{a_1} \tag{5.11}$$

Then V_1 and U_{1S} are calculated from

$$V_1 = W_{1S} \sin 32° \tag{5.12}$$

and

$$U_{1S} = W_{1S} \cos 32° \tag{5.13}$$

It is then possible to determine the shroud diameter, since

$$D_{1S} = \frac{2U_{1S}}{N} \tag{5.14}$$

The hub diameter can be found by applying the mass flow equation (2.5) to the impeller inlet; thus,

$$D_{1H} = \left(D_{1S}^2 - \frac{4\dot{m}}{\pi \rho_1 V_1} \right)^{1/2} \tag{5.15}$$

where the density is determined from the equation of state of a perfect gas, viz.,

$$\rho_1 = \frac{P_1}{RT_1} \tag{5.16}$$

and the static temperature is found from the total temperature using equation (5.10), and the static pressure is found using

$$p_1 = \frac{P_{o1}}{[1 + (\gamma - 1) M_1^2 / 2]^{\gamma/(\gamma - 1)}} \tag{5.17}$$

Referring to Figure 5.3, the fluid angle at the hub is calculated from

$$\beta_{1H} = \tan^{-1} \left(\frac{V_1}{U_{1H}} \right) \tag{5.18}$$

where the vane speed at the hub is given by

$$U_{1H} = \frac{ND_{1H}}{2} \tag{5.19}$$

To determine the impeller diameter, one should follow the procedure used in Example Problem 5.3, viz., first, the dimensional specific speed is calculated, and Table 3 in Appendix A is consulted to determine the highest possible compressor efficiency and the corresponding dimensional specific diameter. Next, the impeller diameter D_2 is calculated from the specific diameter, the tip speed U_2 is determined from the impeller diameter and the energy transfer E is calculated from the ideal energy transfer E_i and the compressor efficiency. The actual tangential velocity component $V_{u2'}$, is calculated from the energy transfer, and the ideal tangential velocity component V_{u2} is calculated from a slip coefficient of from 0.85 to 0.90. Finally, the selection of a flow coefficient in the range of 0.23 to 0.35 permits the calculation of the vane angle and the number of vanes.

The compressor efficiency η_c, in addition to its use in (5.5), can be employed to estimate the impeller efficiency η_I. The ratio χ of impeller losses to compressor losses

$$\chi = \frac{1 - \eta_I}{1 - \eta_c}$$

(5.20)

can be estimated and lies between 0.5 and 0.6. The definition of impeller efficiency

$$\eta_I = \frac{T_{i'} - T_{o1}}{T_{o2} - T_{o1}}$$

(5.21)

can be used to estimate $T_{i'}$, (see Figure 5.1). The latter total temperature corresponds to the total pressure p_{o2}, calculated from

$$\frac{p_{o2}}{p_{o1}} = \left(\frac{T_{i'}}{T_{o1}} \right)^{\gamma/(\gamma - 1)}$$

(5.22)

The static pressure p_2 is then determined from

$$\frac{p_{o2}}{p_2} = [1 + (\gamma - 1)(0.5)M_2^2]^{\gamma/(\gamma - 1)}$$

(5.23)

Table 5.1   Design Parameters for Centrifugal Compressors

Parameter	Source	Recommended range
Flow coefficient	Ferguson	$0.23 < \varphi_2 < 0\,35$
Shroud-tip ratio	Whitfield	$0.5 < D_{1S}/D_2 < 0.7$
Absolute gas angle	Whitfield	$60° < \alpha_{2'} < 70°$
Diffusion ratio	Whitfield	$W_{1S}/W_{2'} < 1.9$

The static pressure p_2 from (5.23) and the static temperature T_2 determined from

$$T_2 = T_{o2} - \frac{V_{2'}^2}{2c_p}$$

$$(5.24)$$

are used to determine density ρ_2 at the impeller exit. Finally, the axial width b_2 of the impeller passage at the periphery may be found from

$$b_2 = \frac{\dot{m}}{2\pi\rho_2 r_2 V_{m2}}$$ *axial width of impeller*

$$(5.25)$$

Ranges of design parameters which are considered optimal by Ferguson (1963) and Whitfield (1990) are presented in Table 5.1. The recommended ranges should be used by the designer to check calculated results for acceptability during or after the design process.

5.3   Diffuser Design

A vaneless diffuser, or empty space, between the leading edges of diffuser vanes and the impeller tip allows some equalization of velocity and a reduction of the exit Mach number. The vaneless portion, which may have a width as large as 6 percent of the impeller diameter, also effects a rise in static pressure. As with the pump, angular momentum rV_u is conserved, and the fluid path is approximately a logarithmic spiral. Diffuser vanes are set with the diffuser axes tangent to the spiral paths and with an angle of

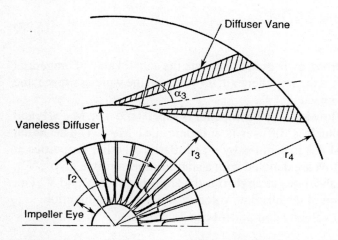

Figure 5.4 Arrangement of diffusers and impeller.

divergence between them not exceeding 12°. The wedge shaped diffuser vanes are depicted in Figure 5.4.

Since the addition of a vaned portion in the diffusion system results in a small-diameter casing, vanes are preferred in instances where size limitations are imposed. On the other hand, a completely vaneless diffuser is more efficient. If vanes are used, then their number should generally be less than the number of impeller vanes to ensure uniformness of flow and high diffuser efficiency in the range of flow coefficient V_{m2}/U_2 recommended in the previous section.

The vaneless diffuser is situated between circles of radii r_2 and r_3. At any radial position r the gas velocity V will have both tangential and radial components. The radial component V_r is the same as the meridional component V_m. The mass flow rate at any r is given by

$$\dot{m} = 2\pi r b \rho V_m \tag{5.26}$$

For constant diffuser width b, the product $\rho r V_m$ is constant, and the continuity equation becomes

$$\rho r V_m = \rho_2 r_2 V_{m2} \tag{5.27}$$

Since angular momentum is conserved in the vaneless space, we can write

$$V_u r = V_{u2'} r_2 \tag{5.28}$$

where the primed subscript is used to indicate the actual value of tangential velocity component at the impeller exit; however, in the vaneless space, the actual velocity is unprimed.

Typically, the flow leaving the impeller is supersonic, i.e., $M_{2'} > 1$, and flow leaving the vaneless diffuser is subsonic, i.e., $M_3 < 1$. The radial position at which $M = 1$ is denoted by r^*; similarly, all other properties at the plane of sonic flow are denoted with a starred superscript, e.g., ρ^*, V_m^*, a^*, T^*, and α^*. The absolute gas angle α is the angle between V and V_r, i.e., between the direction of the absolute velocity and the radial direction.

Since the radial velocity component can be written as

$$V_r = V_m = V \cos \alpha \tag{5.29}$$

the continuity equation becomes

$$\rho r V \cos \alpha = \rho^* V^* r^* \cos \alpha^* \tag{5.30}$$

Similarly, the angular momentum equation is expressed as

$$r V \sin \alpha = r^* V^* \sin \alpha* \tag{5.31}$$

Dividing (5.31) by (5.30), we obtain

$$\frac{\tan \alpha}{\rho} = \frac{\tan \alpha^*}{\rho^*} \tag{5.32}$$

Assuming an isentropic flow in the vaneless region, we find

$$\frac{T}{T^*} = \left(\frac{\rho}{\rho^*}\right)^{\gamma - 1} \tag{5.33}$$

and

$$T = \frac{T_o}{1 + 0.5(\gamma - 1)M^2} \tag{5.34}$$

For M = 1, equation (5.34) becomes

$$T^* = \frac{2T_o}{\gamma + 1} \tag{5.35}$$

Substituting (5.34) and (5.35) into (5.33) yields

$$\frac{\rho^*}{\rho} = \left[\frac{2}{\gamma + 1} \left(1 + \frac{\gamma - 1}{2} M^2 \right) \right]^{1/(\gamma - 1)} \tag{5.36}$$

Substituting (5.36) into (5.32) gives

$$\tan \alpha = \tan \alpha^* \left[\frac{2}{\gamma + 1} \left(1 + \frac{\gamma - 1}{2} M^2 \right) \right]^{-1/(\gamma - 1)} \tag{5.37}$$

The angle α^* is evaluated by substituting $\alpha = \alpha_{2'}$ and $M = M_{2'}$ into (5.37). Equation (5.31) can be rewritten as

$$\frac{r^* \sin \alpha^*}{r \sin \alpha} = \frac{V}{V^*} = \frac{V}{a} \frac{a}{a^*} = M \left(\frac{T}{T^*} \right)^{1/2} \tag{5.38}$$

Sustituting (5.34) and (5.35) into (5.38) yields

$$\frac{r^* \sin \alpha^*}{r \sin \alpha} = M \left[\frac{2}{\gamma + 1} \left(1 + \frac{\gamma - 1}{2} M^2 \right) \right]^{-1/2} \tag{5.39}$$

The radial position r* can be found from (5.39) by substituting $r = r_2$ and $M = M_{2'}$, which are known from impeller calculations. Finally, (5.37) can be used to determine α_3 from a known M_3, and (5.39) can be used to calculate r_3 for known values of M_3 and α_3.

A volute is designed by the same methods outlined in Chapter 4. The volute functions to collect the diffuser's discharge around the 360° periphery and deliver it through a single nozzle to the connecting gas-piping system or to the inlet of the next compressor stage.

5.4 Performance

Typical compressor characteristics are shown in Figure 5.5. Qualitatively, their shape is similar to those of the centrifugal pump, but the sharp fall of the constant-speed curves at higher mass flows is due to choking in some component of the machine. At low flows operation is limited by the phenomenon of surge. Thus, smooth operation occurs on the compressor map at some point between the surge line and the choke line.

The phenomenon of choking is that associated with the attainment of a Mach number of unity. In the stationary passages of the inlet or diffuser, the Mach number is based on the absolute velocity V. Thus for a Mach number of unity, the absolute velocity equals the acoustic speed a, calculated from

$$a = (\gamma RT)^{1/2} \tag{5.40}$$

The temperature at this point is calculated from the total temperature T_0 using the relation

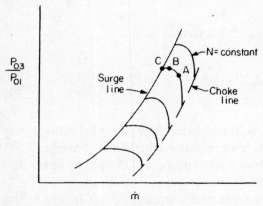

Figure 5.5 Compressor map.

$$T_o = T[1 + (\gamma - 1)(0.5)M^2] \tag{5.41}$$

and setting $M = 1$. Thus

$$T^* = T_o\left(\frac{2}{\gamma + 1}\right) = T_t \tag{5.42}$$

This Mach number is found near the cross section of minimum area, or throat (A_t), so that we can estimate the choking, or maximum, mass flow rate from

$$\dot{m} = A_t P_t\left(\frac{\gamma}{RT_t}\right)^{1/2} \tag{5.43}$$

The pressure P_t at the throat area may be estimated by assuming an isentropic process from the inlet of the stationary component to the throat area. Thus

$$P_t = P_{in}\left(\frac{T_t}{T_{in}}\right)^{\gamma/(\gamma - 1)} \tag{5.44}$$

The process of estimating choked flow rate in the impeller is the same except that relative velocity is substituted for absolute velocity. When the relative Mach number W/a is set equal to unity in the energy equation of the rotor, namely,

$$h_{o1} = h + \frac{W^2}{2} - \frac{U^2}{2} \tag{5.45}$$

we obtain

$$T^* = \left(1 + \frac{U^2}{2c_p T_{o1}}\right)\frac{2T_{o1}}{(\gamma + 1)} = T_t \tag{5.46}$$

Using the isentropic relation between pressure and temperature and substituting into the continuity relation, the mass flow rate at the throat section of the impeller is given by

$$
\dot{m} = A_t P_{o1} \left(\frac{\gamma}{RT_{o1}} \right)^{1/2} \left[\frac{2}{\gamma + 1} \left(1 + \frac{U^2}{2C_p T_{o1}} \right) \right]^{\frac{(\gamma + 1)}{2(\gamma - 1)}}
\tag{5.47}
$$

Thus, it is clear that mass flow for choking in stationary components, given by (5.43), is independent of impeller speed, but that mass flow for choking in the impeller, given by (5.47), actually increases with impeller speed. This is indicated schematically in Figure 5.5.

Referring to Figure 5.5 the point A represents a point of normal operation. An increase in flow resistance in the connected external flow system results in a decrease in V_{m2} at the impeller exit and a corresponding increase in V_{u2}, which results in an increased head or pressure increase. However, the surge phenomenon results when a further increase in external resistance produces a decrease in impeller flow that tends to move the point beyond C, where stall at some point in the impeller leads to change of direction of W_2 and an accompanying decrease in the head (or pressure rise) in the impeller. A temporary flow reversal in the impeller and the ensuing buildup to the original flow condition is known as surging. Surging continues cyclically until the external resistance is removed. It is an unstable and dangerous condition and must be avoided by careful operational planning and system design.

5.5 Examples

Example Problem 5.1

Data from a performance test of a single-stage centrifugal air compressor are the following:

        Measured mass flow rate of air = 2.2 lb_m/s

        Test speed = 60,000 rpm

        Total pressure of air drawn into compressor = 14.7 psia

        Total temperature of air drawn into compressor = 60°F

        Total pressure at compressor discharge = 61.74 psia

Impeller measurements are the following:

Impeller tip diameter = 5.92 inches
Inlet hub diameter = 1.35 inches
Inlet shroud diameter = 3.84 inches
Number of vanes = 33
Vane angle with respect to tangent to wheel = 90°
Calculate the efficiency of this compressor.

Solution: Calculate the tip speed.

$$U_2 = ND_2/2 = (60,000)(\pi/30)(5.92/24) = 1550 \text{ fps}$$

Note that $V_{u2} = U_2$, since $\beta_2 = 90°$. Calculate the slip coefficient using equation (5.7).

$$\mu_s = 1 - 0.63\pi/n_B = 1 - 0.63\pi/33 = 0.94$$

Next determine the tangential velocity component.

$$V_{u2'} = \mu_s V_{u2} = (0.94)(1550) = 1457 \text{ fps}$$

Use equation (5.2) to determine the energy transfer.

$$E = U_2 V_{u2'} = (1550)(1457) = 2,258,350 \text{ ft}^2/\text{s}^2$$

or

$$E = 2,258,350 \text{ ft-lbf/slug} = 70,192 \text{ ft-lb}_f/\text{lb}_m$$

Since air is a diatomic gas, $\gamma = 1.4$. From Table 1 in Appendix F, we find that the molecular weight of air is 28.97. Compute the specific heat of air at constant pressure using the relations for gases, viz., $c_p - c_v = R$ and $\gamma = c_p/c_v$. The resulting equation is

$$c_p = \gamma R/(\gamma - 1)$$

The gas constant is found from

$$R = R_u/M = 1545/28.97 = 53.33 \text{ ft-lb}_f/\text{lb}_m\text{-}°\text{R}$$

Using the above equation the specific heat is

$$c_p = (1.4)(53.33)/(0.4) = 186.66 \text{ ft-lb}_f/\text{lb}_m\text{-}°\text{R}$$

Since no mechanical efficiency is given, we assume that $\eta_m = 0.96$. Calculate the actual total temperature rise in the compressor using equation (5.1); thus,

$$T_{o3} - T_{o1} = E/(c_p\eta_m)$$

$$T_{o3} - T_{o1} = 70,192/(186.66)(0.96) = 392°\text{R}$$

Next determine the isentropic specific work from equation (5.3).

$$E_i = c_p T_{o1}\left[\left(\frac{p_{o3}}{p_{o1}}\right)^{(\gamma-1)/\gamma} - 1\right]$$

$$E_i = (186.66)(520)\left[\left(\frac{61.74}{14.70}\right)^{(0.4/1.4)} - 1\right]$$

$$E_i = 49,197 \text{ ft-lb}_f/\text{lb}_m$$

Calculate the total temperature rise for the isentropic compression.

$$T_i - T_{o1} = E_i/c_p = 49,197/186.66 = 264°\text{R}$$

Finally, calculate the compressor efficiency.

$$\eta_c = \frac{T_i - T_{o1}}{T_{o3} - T_{o1}} = \frac{264}{392} = 0.673$$

Example Problem 5.2

Determine the compressor efficiency for the compressor described in Example Problem 5.1 using Table 3 in Appendix A.

Solution: Determine the density at the inlet. Note that the eye velocity V_1 must be found iteratively. Assume $V_1 = 447$ fps. Calculate T_1.

$$T_1 = T_{o1} - \frac{V_1^2}{2c_p}$$

$$c_p = 186.66(32.174) = 6006 \text{ ft-lb}_f/\text{slug-}°R$$

$$T_1 = 520 - \frac{(447)^2}{2(6006)} = 503°R$$

$$p_1 = p_{o1}\left(\frac{T_1}{T_{o1}}\right)^{\gamma/(\gamma-1)} = 14.7\left(\frac{503}{520}\right)^{3.5} = 13.09 \text{ psia}$$

$$\rho_1 = \frac{p_1}{RT_1} = \frac{13.09(144)}{53.33(503)} = 0.070 \text{ lb}_m/\text{ft}^3$$

Calculate flow area at eye.

$$A_1 = \frac{\pi}{4}(D_{1S}^2 - D_{1H}^2)$$

$$A_1 = \frac{\pi[(3.84)^2 - (1.35)^2]}{4(144)} = 0.0705 \text{ sq ft}$$

Check mass flow rate to verify assumed V_1.

$$\dot{m} = \rho_1 V_1 A_1 = (0.070)(447)(0.0705) = 2.2 \text{ lb/s}$$

Calculate inlet volume flow rate.

$$Q_1 = \frac{\dot{m}}{\rho_1} = \frac{2.2}{0.070} = 31.4 \text{ cfs}$$

Calculate output head H.

$$H = \frac{E_i}{g} = \frac{49,197(32.174)}{32.174} = 49,197 \text{ ft}$$

Calculate dimensional specific speed.

$$N_s = \frac{NQ_1^{1/2}}{(H)^{3/4}} = \frac{60,000(31.4)^{1/2}}{(49,197)^{3/4}} = 102$$

Calculate dimensional specific diameter.

$$D_s = \frac{D_2 H^{1/4}}{Q^{1/2}} = \frac{(5.92/12)(49,197)^{1/4}}{(31.4)^{1/2}} = 1.31$$

Interpolate in Table 3 in Appendix A to find efficiency.

$$\eta_c = 0.70 \text{ for } D_s = 1.3 \text{ and } N_s = 102.$$

Note: This is slightly higher than the 0.673 calculated previously. If $\eta_m = 0.99$ had been assumed in Example Problem 1, the agreement would have been perfect.

Example Problem 5.3

A single-stage centrifugal compressor draws in 6.9 lb/s of air at a total pressure of 14.2 psia and a total temperature of 550°R. It discharges the air at a total pressure of 59.64 psia. The compressor runs at 41,700 rpm. Find the basic dimensions of the impeller.

Solution: Assume a slightly supersonic relative Mach number at the inlet shroud. Although undesirable to create a region of supersonic flow in the inlet, it is necessary to handle the large mass flow. Choose $M_{R1S} = 1.2$ and $\beta_{1S} = 32°$. Calculate the inlet density and the volume flow rate using equations (5.8)–(5.12).

$$M_1 = M_{R1S} \sin \beta_{1S} = 1.2 \sin 32° = 0.636$$

$$T_1 = \frac{550}{1 + 0.2(0.636)^2} = 509°R$$

$$a_1 = [(1.4)(1716)(509)]^{1/2} = 1106 \text{ fps}$$

$$V_1 = (0.636)(1106) = 703 \text{ fps}$$

$$\rho_1 = \rho_{o1}\left(\frac{T_1}{T_{o1}}\right)^{1/(\gamma-1)}$$

$$\rho_1 = .0697\left(\frac{509}{550}\right)^{2.5} = 0.0574 \text{ lb/cu ft}$$

$$Q_1 = \frac{\dot{m}}{\rho_1} = \frac{6.9}{0.0574} = 120 \text{ cfs}$$

Calculate output head H.

$$E_i = 6006(550)\left[\left(\frac{59.64}{14.20}\right)^{1/3.5} - 1\right] = 1,674,291 \text{ ft}^2/\text{s}^2$$

$$H = \frac{E_i}{g} = \frac{1,674,291}{32.174} = 52038 \text{ ft}$$

Calculate the dimensional specific speed.

$$N_s = \frac{NQ_1^{1/2}}{H^{3/4}} = \frac{41,700(120)^{1/2}}{(52038)^{3/4}} = 132.5$$

Referring to Table 3 in Appendix A for highest efficiency, $D_s = 1.18$ and

$$D_2 = \frac{D_sQ_1^{1/2}}{H^{1/4}} = \frac{(1.18)(120)^{1/2}}{(52,038)^{1/4}} = 0.855 \text{ ft} = 10.2 \text{ inches}$$

The corresponding tip speed is

$$U_2 = (4366.8)(10.2/24) = 1856 \text{ fps}$$

The average impeller temperature t_{av} is estimated to be 200°F. The recommended maximum tip speed for this metal temperature is

$$U_{2max} = 2000 - t_{av} = 2000 - 200 = 1800 \text{ fps}$$

Choose a design tip diameter of 9.6 inches.

$$U_2 = (4366.8)(9.6/24) = 1747 \text{ fps}$$

The corresponding D_s is 1.103. From Table 3 in Appendix A,

$$\eta_c = 0.723$$

The energy transfer is

$$E = \frac{\eta_m E_i}{\eta_c} = \frac{0.96(1,674,291)}{0.723} = 2,223,125 \text{ ft}^2/\text{s}^2$$

The actual tangential velocity component is found from (5.2).

$$V_{u2'} = \frac{2,223,125}{1747} = 1273 \text{ fps}$$

Choose a slip coefficient of 0.9. Then

$$V_{u2} = \frac{1273}{0.9} = 1414 \text{ fps}$$

and

$$W_{u2} = U_2 - V_{u2} = 1747 - 1414 = 333 \text{ fps}$$

Choose the flow coefficient of 0.30. This is the middle of the design range.

$$W_{m2} = \varphi_2 U_2 = (0.3)(1747) = 524 \text{ fps}$$

The vane angle is

$$\beta_2 = \tan^{-1} \frac{W_{m2}}{W_{u2}} = \tan^{-1}\left(\frac{524}{333}\right) = 57.6°$$

Solve the slip equation (5.7) for the number of vanes.

$$0.9 = 1 - \frac{0.63\pi}{n_B}\left(\frac{1}{1 - 0.3 \cot 57.6}\right)$$

$$n_B = 24$$

Estimate the impeller efficiency using a loss ratio of 0.55. From (5.20), we have

$$\eta_I = 1 - 0.55(1 - 0.723) = 0.848$$

Equation (5.1) can be used to find T_{o3}.

$$T_{o3} = 550 + \frac{2{,}223{,}125}{6006(0.96)} = 936°R$$

Note: $T_{o2} = T_{o3} = 936°R$. Use equation (5.21) to find $T_{i'}$.

$$T_{i'} = 550 + (0.848)(936 - 550) = 877°R$$

The impeller pressure ratio is found from equation (5.22).

$$\frac{P_{o2}}{P_{o1}} = \left(\frac{887}{550}\right)^{3.5} = 5.12$$

The density based on total properties is

$$\rho_{o2} = \frac{P_{o2}}{RT_{o2}} = \frac{(14.2)(5.12)(144)}{(53.33)(936)} = 0.21 \text{ lb/cu ft}$$

The gas temperature at the impeller exit is

$$T_2 = T_{o2} - \frac{V_{2'}^2}{2c_p}$$

$$V_{2'}^2 = (1273)^2 + (524)^2$$

$$T_2 = 778°R$$

The tip density is

$$\rho_2 = \rho_{o2}\left(\frac{T_2}{T_{o2}}\right)^{1/(\gamma-1)}$$

$$\rho_2 = 0.21\left(\frac{778}{936}\right)^{2.5} = 0.132 \text{lb/cu ft}$$

Determine the tip width using equation (5.25).

$$b_2 = \frac{6.9(12)}{\pi(9.6/12)(0.132)(524)} = 0.48 \text{ inch}$$

Determine the inlet dimensions from equations (5.13)–(5.15).

$$U_{1S} = (1.2)(1106)\cos 32° = 1126 \text{ fps}$$

$$D_{1S} = \frac{24(1126)}{4366.8} = 6.2 \text{ inches}$$

$$D_{1H} = \left[(6.2)^2 - \frac{4(6.9)(144)}{\pi(0.0574)(703)}\right]^{1/2} = 2.7 \text{ inches}$$

Note: $D_{1S}/D_2 = 0.65$, which is within the design limits. Calculate the absolute gas angle.

$$\alpha_{2'} = \tan^{-1}\left(\frac{V_{u2'}}{W_{m2}}\right) = \tan^{-1}\left(\frac{1273}{524}\right) = 67.6° < 70°$$

The absolute gas angle is inside the acceptable range. Calculate the diffusion ratio.

$$W_{1S} = M_{R1S}a_1 = (1.2)(1106) = 1327 \text{ fps}$$

$$W_{u2'} = U_2 - V_{u2'} = 1747 - 1273 = 474 \text{ fps}$$

$$W_{2'} = [W_{u2'}^2 + W_{m2}^2]^{1/2}$$

$$W_{2'} = [(474)^2 + (524)^2]^{1/2} = 707 \text{ fps}$$

$$\frac{W_{1S}}{W_2} = \frac{1327}{707} = 1.877 < 1.9$$

The diffusion ratio is high but in the acceptable range.

References

Ferguson, T. B. 1963. *The Centrifugal Compressor Stage*. Butterworths, London.
Shepherd, D.G. 1956. *Principles of Turbomachinery*. Macmillan, New York.
Whitfield. A. 1990. *Journal of Power and Energy*. Institution of Mechanical Engineers. U.K.

Problems

5.1. Air enters a centrifugal compressor at 1 atm, 58°F, and $V_1 = 328$ ft/s. At the impeller exit $\beta_{2'} = 63.4°$, $V_{m2} = 394$ ft/s, and $U_2 = 1640$ ft/s. The mass flow rate is 5.5 lb/s, the mechanical efficiency is 95 percent, and the compressor efficiency is 80 percent. Determine the ratio of total pressures at outlet and inlet and the power required to drive the machine.

5.2. Design a single-stage centrifugal compressor which will handle 2.2 lb/s of air at a pressure ratio of 4.2:1. Use 33 radial blades with an appropriate inducer. Assume that $\eta_c = 0.70$. The machine is to operate at 60,000 rpm and to supply air to the combustion chamber of a turbojet engine. The basic design parameters required are the following:

   a) hub diameter
   b) shroud diameter at impeller inlet
   c) shroud diameter at impeller exit

  d) impeller inlet vane angle
  e) vane width at impeller exit
  f) velocity triangle at impeller inlet (shroud diameter)
  g) velocity triangle at impeller exit
The compressor is to have no inlet guide vanes and is to draw in ambient air at 14.7 psia and 520°R. Assume $\eta_m = 0.99$

5.3. Derive equation (5.3) from equation (5.1). Hint: $h'_{o2} = h_{o3}$.

5.4. Consider the high-speed flow of a diatomic gas at the inlet of the impeller of a centrifugal compressor. Show that

$$\frac{M_{R1S}^3 \cos^2 \beta_{1S} \sin \beta_{1S}}{[1 + (0.2M_{R1S}^2) \sin^2 \beta_{1S}]^4} = \frac{C_1 N^2 \dot{m}}{P_{o1}(T_{o1})^{1/2}(1 - k^2)}$$

where C_1 is a constant and k is the hub-tip ratio at the impeller inlet.

5.5. Plot the left hand side of the equation of Problem 5.4 as a function of β_{1S} with M_{R1S} as parameter. Use subsonic values of the relative Mach number, e.g., 0.5, 0.6, 0.7, 0.8, and 0.9. Show that the maximum ordinate for each curve occurs at the same value of β_{1S} (32°).

5.6. Test data resulting from the test of a single-stage centrifugal air compressor are the following:

  $P_{o1} = 14.5$ psia
  $T_{o1} = 58°F$
  $p_{o3} = 60.9$ psia
  $T_{o3} = 450°F$ $\dot{m} = 1 \frac{Lb}{s}$

Assume $\eta_m = 0.96$ and calculate:

  a) the power required to drive the compressor
  b) the compressor efficiency
  c) the energy transfer

5.7. A single-stage centrifugal compressor operating at a speed of 15,000 rpm compresses air from an inlet total pressure of 14.7 psia to a discharge total pressure of 24.7 psia. The compressor is driven by an 80 horsepower motor. The mechanical efficiency of the compressor is 0.96, and the inlet total temperature is 528°R. The volume flow rate of air handled, measured at inlet conditions, is 1350 cubic feet per minute(cfm). Find:

  a) mass flow rate of air in pounds per second
  b) energy transfer for isentropic compression in Btu/lb
  c) output head in feet of air

d) actual energy transfer in Btu/lb

e) input head in feet of air

f) compressor efficiency

5.8. The speed of the compressor described in Problem 5.7 is increased until the discharge total pressure is 29.5 psia. Find:

    a) new output head in feet of air

    b) new speed in rpm

    c) new volume flow rate at inlet conditions in cfm

5.9. Air enters a single-stage centrifugal compressor with a total pressure of 1.013 bar and a total temperature of 288°K. The axial velocity at the impeller eye is 100 m/s. The rotor tip speed is 500 m/s, $V_{2'}$ is 456 m/s, and $V_{u2'}$ is 440 m/s. The impeller efficiency is 0.9 and that of the compressor stage is 0.8. The mechanical efficiency is 0.98. Find the static pressure in bar at the impeller outlet.

5.10. A single stage centrifugal compressor receives air at a stagnation pressure of 1.013 bar and a stagnation temperature of 288°K. The rotor tip speed is 500 m/s, the velocity at the impeller eye is 100 m/s and $V_{2'}$ is 456 m/s. The energy transfer in the impeller is 220 kilo joules/kg. The slip coefficient is 0.9. The impeller efficiency is 0.9, and the compressor efficiency is 0.8. The mechanical efficiency is 0.98. Find:

    a) tip vane angle

    b) T_{o2}

    c) T_2

    d) P_{o2}

    e) P_2

5.11. Air enters a single-stage centrifugal compressor at 1 atm and 58°F. The entering flow is axial with $V_1 = 328$ fps. At the impeller tip $\beta_{2'} = 63.4°$, $V_{m2} = 394$ fps and $U_2 = 1640$ fps. The mass rate of flow is 5.5 lb/sec. The mechanical efficiency is 95 percent, and the compressor efficiency is 80 percent. Find:

    a) $V_{u2'}$

    b) P_{o3}/P_{o1}

    c) brake power

5.12. A performance test of single-stage centrifugal air compressor yielded the following data: $P_{o3} = 44$ psia, $T_{o3} = 770°R$, $P_2 = 28$ psia, $N = 45,960$ rpm, $P_{o1} = 14.5$ psia, $T_{o1} = 520°R$ and the mass flow of air was 1.4 lb/s. The impeller had 17 vanes, and its principal dimensions are $D_2 = 6.5$ inches and $b_2 = .394$ inch. Assuming a mechanical efficiency of 0.96, find:

    a) compressor efficiency
    b) impeller efficiency
    c) fraction of the overall loss occurring in the impeller

5.13. A single-stage air compressor wlth 20 radial vanes ($\beta_2 = 90°$) has an efficiency of 78 percent, an impeller efficiency of 89 percent, and a mechanical efficiency of 96 percent. The compressor handles 19.8 lb/s of air and turns at 17,400 rpm. $P_{o1} = 15.96$ psia, $T_{o1} = 531°R$, $V_{u2'} = 1345$ fps and $V_{m2} = 469$ fps. Find
    a) D_2
    b) E
    c) P_{o3}
    d) P_{o2}
    e) P_2
    f) b_2

5.14. Test data for a single-stage centrifugal air compressor are the following: $D_2 = 18$ inches, vane angle at impeller tip $\beta_2 = 60°$, N = 18,000 rpm, mass rate of flow = 19 lb/s, $T_{o1} = 520°R$, $T_{o2} = 778°R$, $P_{o1} = 14.1$ psia and $P_{o3} = 44.9$ psia. Assume that the slip coefficient is 0.90, and the mechanical efficiency is 96 percent. Find:
    a) compressor efficiency
    b) flow coefficient W_{m2}/U_2
    c) shaft power required in horse power

5.15. Design data for a single-stage centrifugal air compressor are the following: $D_2 = 19.7$ inches, N = 16,200 rpm, $\dot{m} = 35$ lb/s, $T_{o1} = 58°F$, $P_{o1} = 14.6$ psia, impeller efficiency = 90 percent, mechanical efficiency = 96 percent, $\beta_2 = 90°$, $n_B = 20$, $b_2 = 1.97$ inches and radial gap of vaneless space, $r_3 - r_2 = 1.6$ inches. Find:
    a) P_{o2}
    b) T_{o2}
    c) M_2
    d) M_3
    e) α_3. This is the angle at which the diffuser vanes should be set.

5.16. A single-stage centrifugal compressor is to handle 2.25 lbs/sec of air at a total pressure ratio of 4.15. The inlet air has a total pressure of 14.5 psia and a total temperature of 65°F. At the impeller tip the 16 full blades and 16 splitters have a blade angle of 90° with the tangential direction. The impeller is to turn at 60,000 rpm. Find:
    a) D_{1S}
    b) D_{1H}

c) D_2

d) b_2

5.17. The first stage of a centrifugal air compressor is to handle 6.938 lb/s of air at 14.168 psia and 560°R (total properties). The impeller is to rotate at 41,730 rpm and to produce a pressure ratio of 4.21. Determine the basic dimensions of the impeller, vane angles, and number of vanes.

5.18. Design a vaneless diffuser to match the impeller of Problem 5.2. Assume that $D_2 = 5.92$ inches, $b_2 = 0.281$ inch, $T_2 = 700°R$, $V_{m2} = 463$ fps, $M_{2'} = 1.172$, and $\alpha_{2'} = 72.3°$. Calculate α_3 and r_3 at the diffuser exit, if $M_3 = 0.80$. Assume that b is constant in the vaneless space.

Symbols for Chapter 5

a	local acoustic speed in gas
a_1	acoustic speed at impeller inlet
b	width of vaneless diffuser
b_2	width of vane at $r = r_2$ in compressor impeller
c_p	specific heat at constant pressure
D_{1S}	shroud diameter at impeller inlet
D_{1H}	hub diameter at impeller inlet
D_2	impeller tip diameter
D_s	specific diameter
E	energy transfer from impeller to fluid
E_i	ideal energy transfer for isentropic compression from p_{o1} to $p_{o3} = gH$
g	gravitational acceleration
H	output head $= E_i/g$
h_{o1}	total enthalpy of gas entering impeller
h_{o2}	total enthalpy of gas leaving impeller
h_{o3}	total enthalpy of gas leaving diffuser
k	hub-tip ratio at impeller inlet $= D_{1H}/D_{1S}$
M	molecular weight of gas
M	Mach number
M_1	absolute Mach number at impeller inlet $= V_1/a_1$
M_R	relative Mach number $= W/a$
M_{R1S}	relative Mach number $= W_{1S}/a_1$
$\dot{m}$	mass flow rate of gas discharged from compressor
$\dot{m}_L$	mass flow rate of gas leaked from the high to the low pressure regions outside of impeller

N	rotor speed
N_s	specific speed
n_B	number of vanes (or blades) in the impeller
P	power to impeller shaft
p_1	static pressure at impeller inlet
p_{o1}	total pressure of gas at stage inlet
p_{o2}	total pressure of gas at impeller outlet
p_{o3}	total pressure of gas at diffuser outlet
Q_1	volume flow rate based on gas density at impeller inlet
r	radial position measured from axis of rotation
r_2	radial position at tip of impeller $= D_2/2$
r_3	radial position at exit from vaneless diffuser
r_4	radial position at exit from vaned diffuser
R	gas constant $= R_u/M$ molecular weight
R_u	universal gas constant
T	gas temperature
T_o	total temperature of gas
T^*	gas temperature where $M = 1$ (stationary frame) or $M_R = 1$ (moving frame)
T_1	static temperature at impeller inlet
T_{o1}	total temperature of gas entering impeller
T_{o2}	total temperature of gas leaving impeller
T_{o3}	total temperature of gas leaving diffuser
T_i	total temperature of gas at the end of an isentropic compression from p_{o1} to p_{o3}
U	vane speed at any r
U_1	impeller speed at leading edge of vane
U_{1S}	U_1 at shroud diameter
U_{1H}	U_1 at hub diameter
U_2	impeller speed at vane tip
V_1	absolute velocity at leading edge of impeller vane
V_{u1}	tangential component of V_1
V_2	absolute velocity of gas leaving the vane of an impeller with an infinite number of vanes
$V_{2'}$	absolute velocity of gas leaving the vane of an impeller with a finite number of vanes
V_m	meridional component of V at any $r = V_r$
V_{m2}	meridional component of V_2 or $V_{2'} = W_{m2}$
V_r	radial component of V at any $r = V_m$

V_u tangential component of V at any r

V_{u2} tangential component of V_2

$V_{u2'}$ tangential component of $V_{2'}$

W   velocity relative to moving vane

W_1 relative velocity at leading edge of impeller vane

W_2 relative velocity of gas leaving the vane of an impeller with an infinite number of vanes

$W_{2'}$ relative velocity of gas leaving the vane of an impeller with a finite number of vanes

W_{m2} meridional component of W_2, $W_{2'}$, V_2, or $V_{2'}$

W_{1S} relative velocity of gas entering impeller at shroud

α absolute gas angle = angle between V and V_r

α^* absolute gas angle where M = 1

$\alpha_{2'}$ $\tan^{-1}(V_{u2'}/W_{m2})$; absolute gas angle at $r = r_2$

α_3 absolute gas angle at $r = r_3$

β_1 angle between W_1 and U_1

β_{1S} angle between W_{1s} and U_{1s}

β_2 angle between W_2 and U_2; also the vane angle at trailing edge of vane

$\beta_{2'}$ angle between $W_{2'}$ and U_2; actual fluid angle

γ ratio of specific heats

η_c stage efficiency

η_m mechanical efficiency

μ_s slip coefficient

ρ gas density

ρ^* gas density where M = 1

ρ_1 gas density at impeller inlet

ρ_2 gas density at impeller exit

φ_2 flow coefficient at impeller exit = W_{m2}/U_2

6

Axial-Flow Compressors, Pumps, and Fans

6.1 Introduction

Originally a very inefficient machine, the axia ɔw compressor was not used to compress air in the gas turbine power plant. However, the devel-opment of the science of aerodynamics, which accompanied the dev opment of high performance aircraft, made possible its present use in gas-turbines. Now a highly efficient machine, it must be studied and understood thoroughly by engineers.

This machine resembles the axial-flow steam or gas turbine in general appearance. Usually multistage, one c erves rows of blades on a single shaft with blade length varying monotonically as the shaft is traversed. The difference is, of course, that the blades are shorter at the outlet end of the compressor, whereas the turbine receives gas or vapor on short blades and exhausts it from long blades. A close look at the blades shows that the compressor blade deflects the fluid through only a fraction of the angle that the turbine blade does. This point is illustrated by Figure 6.1. Figure 6.1 also indicates that the concave side of the blade moves ahead of the convex side; the reverse is true of the turbine blade. Clearly, the fluid receives energy from the compressor blade and gives up energy to the turbine blade. Aerodynamic analysis must be carried out for compressor blades, since flow in the boundary layer encounters an adverse pressure gradient, which may lead to separation, stall, and the consequent surge phenomenon discussec in connection with centrifugal machines. To avoid separation the pressure

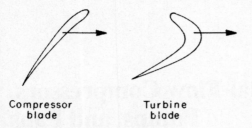

Compressor
blade

Turbine
blade

Figure 6.1 Blade comparison.

rise must be small for each stage, in contrast with the very large pressure drops found in turbine stages. Typically, about one-half of the enthalpy rise occurs in the rotor and one-half in the stator.

The approach to compressor stage design is the same as that used for axial-flow pumps and fans, except that the compressibility (density change) of the gas must be considered in the overall process of multistage machines. Fortunately, an abundance of theory exists, and many blade shapes have been tested in cascade tunnels, so that the designer has a large stock of data to draw on in his or her considerations.

Axial-flow pumps and fans move liquids and gases without significant effect on their density. They are like propellers in that power is supplied to produce axial motion of the fluid, but they are different in that the fluid being moved is enclosed by a casing. Like propellers, the vanes have small curvature and cause little deflection of the relative velocity vector of a fluid particle as it migrates through the moving passages.

Generally, the vanes have shapes, or profiles, like that of an airfoil: they are thin, streamlined, and cambered (Figure 6.2). The relative velocity W_1 approaches the vane at an angle α (the angle of attack) to the chord line. The exiting fluid with relative velocity W_2 has been deflected slightly,

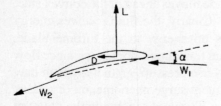

Figure 6.2 Blade profile.

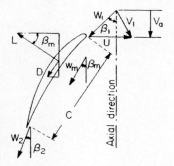

Figure 6.3 Blade motion.

and the change of momentum results in a lift force L perpendicular to the mean direction of W_1 and W_2, i.e., perpendicular to a mean relative velocity W_m.

The lift force L is primarily responsible for the transfer of energy, and the drag force D, which is directed parallel to W_m, is strongly associated with blade losses. Lift is maximized by setting the blades at high angle of attack, but stall occurs if the angle is too high. Such characteristics of blades are determined in a wind tunnel using a representative set of blades arranged in series, known as a cascade. This kind of experimentation provides information not only about optimum incidence, but also about optimum spacing for maximum lift and minimum drag.

The axial-flow fan, pump, or compressor blade moves in a direction opposite to that of the blade in an axial-flow turbine; the concave side comes first. In Figure 6.3 blade motion is to the right. The tangential component F_{Bu} of the blade force can be obtained in terms of the angle β_m that W_m makes with the axial direction. Referring to Figure 6.4, it is clear that

$$F_{Bu} = L \cos \beta_m + D \sin \beta_m \qquad (6.1)$$

The rate of energy transfer E is given by

$$\dot{E} = U(L \cos \beta_m + D \sin \beta_m) \qquad (6.2)$$

and the energy transfer per unit mass is expressed as

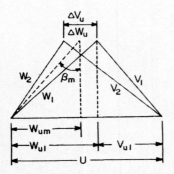

Figure 6.4 Control volume for a cascade blade.

$$E = U(V_{u2} - V_{u1}) \tag{6.3}$$

in which U is the same at the inlet and exit planes, since the flow ideally contains no radial components of velocity. An alternative expression for $\dot{E}$ is obtained by multiplying (6.3) by mass flow rate $\dot{m}$, given by

$$\dot{m} = \rho V_a s \tag{6.4}$$

where V_a is the axial component of absolute velocity, and s is the spacing between two adjacent blades in a row. It is evident from (6.4) that $\dot{m}$ is mass flow rate per blade per unit length of blade. The blade force equation (6.1), and the blade power equation (6.2), should also be interpreted as force and power, respectively, per blade per unit length of blade.

Equating (6.2) with (6.3) times (6.4) and nondimensionalizing, we obtain

$$c_L = 2 \frac{\cos^2 \beta_1}{\cos \beta_m} \frac{s}{c} (\tan \beta_1 - \tan \beta_2) \tag{6.5}$$

where, as is evident from Figure 6.5,

$$\Delta V_u = V_{u2} - V_{u1} = W_{u1} - W_{u2} = \Delta W_u \tag{6.6}$$

In getting (6.5) we have defined the lift coefficient c_L as

$$c_L = \frac{L}{\frac{1}{2}\rho W_1^2 c} \tag{6.7}$$

where c is the chord, the length of a line drawn between the leading edge and the trailing edge of the blade (Figure 6.3). Also, in forming (6.5) we have omitted the drag term that appears in (6.1) on the grounds that $D \ll L$. The relationship expressed by (6.5) links the force produced by a blade with the flow angle and the nondimensional spacing of blades, or pitch-chord ratio s/c. For the ideal cascade, c_L can be predicted from (6.5). The presence of boundary layers and nonuniform flow deflection in actual cascades leads to experimentally determined relationships between these variables.

6.2 Stage Pressure Rise

The first stage of an axial-flow compressor, fan or pump may include inlet guide vanes, which deflect the fluid from an axial path by the angle α_1 and increase its velocity from V_a to V_1 (Figure 6.3), so that the fluid enters the moving rotor blades at the proper angle. Figure 6.5 shows that $V_2 > V_1$, which means that the kinetic energy of the fluid has been increased by the action of the rotor blades. In a single-stage machine, the stator blades may turn the fluid back to the axial direction. In multistage machines, the stator blades redirect the fluid to its original direction, so that the fluid leaves the stage at absolute fluid angle $\alpha_3 = \alpha_1$ with absolute velocity

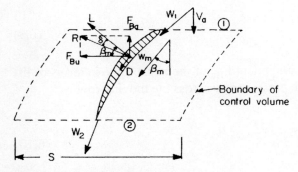

Figure 6.5 Velocity diagrams.

$V_3 = V_1$. Equation (2.25), applied to an axial-flow compressor, fan or pump, shows that $h_2 > h_1$, since $W_1 > W_2$. For liquids or gases, the enthalpy rise implies a corresponding pressure rise in the rotor, and, for gases, temperature increase is also indicated. Equation (2.9), applied to the stator or diffuser of an axial-flow stage, simplifies to

$$h_2 + \frac{V_2^2}{2} = h_3 + \frac{V_3^2}{2}$$

$$(6.8)$$

and shows that $h_3 > h_2$ and $p_3 > p_2$; for compressors, however, $T_3 > T_2$ is also indicated. Thus, a complete stage effects an acceleration in the rotor and a deceleration in the stator, both of which are accompanied by a pressure rise.

Compressors, pumps, and fans have solidities ranging from as low as 0.1 for two-bladed pumps or fans up to 1.5 for some compressors. Axial-flow machines of low solidity are typically machines of high specific speed, as is shown in Table 6.2. Eck (1973) shows that optimum solidity, defined as the optimum ratio of blade chord length to blade spacing, is proportional to $\tan \beta_1 - \tan \beta_2$ which implies that machines of low energy transfer, or head, require low solidities, and low head correlates with high specific speed. Increasing the number of vanes increases guidance and thereby increases head, but friction losses also are increased. The optimum solidity is determined by test.

Let us consider a control volume, surrounding a single moving blade, of width S and of unit height along the blade, as shown in Figure 6.4. Assuming no change in the axial component of fluid velocity from inlet to outlet, we can write the force equilibrium equation as

$$(P_2 - P_1)S - F_{Ba} = 0$$

$$(6.9)$$

Expressing the axial component F_{Ba} of the blade force in terms of lift and drag, we obtain for the pressure rise across the moving row

$$(P_2 - P_1)_{rotor} = \frac{L}{S} \sin \beta_m - \frac{D}{S} \cos \beta_m$$

$$(6.10)$$

In nondimensional form, (6.10) becomes

$$\frac{(P_2 - P_1)_r}{\frac{1}{2}\rho W_1^2} = \frac{c}{s} (c_L \sin \beta_m - c_D \cos \beta_m)$$

(6.11)

Thus (6.11) allows prediction of pressure rise in terms of aerodynamic coefficients and relative velocities. A similar relation can be obtained for the diffuser section of the stage using the same method.

An alternate expression for pressure rise across the moving row is obtained by expressing the axial component F_{Ba} in terms of the tangential component F_{Bu}. Thus,

$$F_{Ba} = F_{Bu} \tan (\beta_m - \delta)$$

(6.12)

Since it is small, the tangent of δ may be approximated by δ, which represents the ratio of drag to lift, or c_D/c_L. Using a trigonometric identity and noting from Figure 6.4 that

$$\tan \beta_m = \frac{|W_{um}|}{V_a} = \frac{R}{\varphi}$$

(6.13)

we substitute (6.12) and (6.13) into (6.9) to obtain

$$(P_2 - P_1)_r = \rho U^2 \varphi \Lambda_B \frac{R - \varphi\delta}{\varphi + \delta R}$$

(6.14)

in which the component F_{Bu} is replaced by using the change in tangential momentum flow, namely,

$$F_{Bu} = \rho V_A S(V_{u2} - V_{u1}) = \rho U^2 S \Lambda_B \varphi$$

(6.15)

where Λ_B is the blade loading coefficient $\Delta V_u/U$. A similar derivation may be made for the diffuser blading to obtain the pressure rise in the stationary part of the stage. Addition of the two leads to the equation for overall stage pressure rise:

$$\Delta p = \rho U^2 \varphi \Lambda_B \left[\frac{R - \varphi\delta}{\varphi + \delta R} + \frac{1 - R - \varphi\delta}{\varphi + \delta(1 - R)} \right]$$

(6.16)

The above relation is useful in estimating stage efficiency, which is defined as the ratio of pressure rise with blade drag accounted for to that with frictionless blades. Elimination of drag means that the terms in (6.16) that involve δ vanish. Thus the stage efficiency η is given by

$$\eta_s = \frac{\Delta p}{\Delta p_{ideal}} = \varphi \left[\frac{R - \varphi\delta}{\varphi + \delta R} + \frac{1 - R - \varphi\delta}{\varphi + \delta(1 - R)} \right] \tag{6.17}$$

The drag-lift ratio δ used in (6.17) must be modified to account for several additional losses. These are considered in the next section.

6.3   Losses

Boundary layers on the surfaces of blades, whether moving or stationary, mark regions of high shear stress, and the resultant of the viscous forces produced at the surface of the blade is the drag force. In addition to resisting blade movement, viscous forces retard fluid in the stationary passages and result in total pressure losses. The thickness of the boundary layers on the blade surfaces deflects the main flow and thus changes the effective blade shape. The effect of the increasing pressure in the flow direction slows down the fluid in the boundary layer and promotes separation of the boundary layer from the blade surfaces concomitantly creating regions of reversed flow. In the latter case, the effective blade shape is drastically distorted, and the flow direction is severely modified.

Besides boundary-layer formation on the blades, layers are also formed on the inner and outer surfaces of the annular-flow passage, the cylindrical surfaces at the hub and tip radii. Since the flow actually takes place in the rectangular passage bounded by blades on two sides and by walls of the annulus on the other two, it is expected that losses will depend on the ratio of blade spacing to blade height. Empirically, it has been found that the drag produced by these surfaces is correctly reflected by the relation

$$c_{D'} = 0.02 \frac{S}{h} \tag{6.18}$$

where h represents blade height and $c_{D'}$ is the increment to be added to the previous drag coefficient to account for annulus losses.

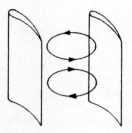

Figure 6.6 Secondary flow in blade passages.

The velocity variation due to boundary layers on the blades and walls of the annulus, coupled with the curvature of the blade surfaces, results in an additional loss. Secondary currents are set up in a plane transverse to the flow, as is indicated in Figure 6.6. Dissipation of the energy of these secondary currents takes place in the blade passage and in the wakes behind the trailing edges via vortices spawned by the interaction of neighboring secondary flow cells as they leave the blade. Because the trailing vortices are similar to wing vortices, it is expected that the corresponding drag is proportional to the square of the lift coefficient. The recommended equation for drag coefficient is then

$$c_{D''} = 0.018 c_L^2 \qquad\qquad (6.19)$$

The difference in pressure on the two sides of the moving blades results in a leakage of fluid around the tip, i.e., through the narrow passage formed between the blade tip and the casing. This loss is accounted for by the empirical formula

$$c_{D'''} = 0.29\, \frac{c_T}{h}\, c_L^{3/2} \qquad\qquad (6.20)$$

where c_T is the tip clearance.

To obtain a more realistic value for the stage efficiency using (6.17), we can artificially increase the blade drag force by an amount proportional to the sum of $c_{D'}$, $c_{D''}$, and $c_{D'''}$. We then substitute for δ in (6.17) using the expression

$$\delta = \frac{c_D + c_{D'} + c_{D''} + c_{D'''}}{c_L} \tag{6.21}$$

6.4 Pump Design

Axial flow pumps are used for specific speeds above approximately 3, with centrifugal pumps occupying the range below 2 and mixed-flow pumps filling the gap between the two. They are then machines of low head, high capacity, and a single stage. They require several well-finished rotor vanes of airfoil section, as shown in Figure 6.2.

As a starting point in the design of an axial-flow pump we can use that part of a Cordier diagram (Figure 3.3) for which $N_s > 3$. The relationship between specific speed N_s and specific diameter D_s is given approximately by

$$D_s = \frac{2.95}{N_s^{0.485}} \tag{6.22}$$

Calculating N_s from specified values of N, Q, and H, we can arrive at D_s from (6.22). This value of specific diameter is used to compute a rotor tip diameter D_t in the following manner:

$$D_t = \frac{D_s Q^{\frac{1}{2}}}{(gH)^{\frac{1}{4}}} \tag{6.23}$$

By selecting a suitable hub-tip ratio, i.e., blade-root diameter D_r divided by blade-tip diameter D_t, we are able to compute the hub diameter, which is a synonymous expression for root diameter. Generally, axial-flow pumps have hub-tip ratios in the range 0.3 to 0.7. The graphical display of the relationship between D_r/D_t, N_s, and solidity c/s, based on current practice, is given by Stepanoff (1957) and may be approximated by the relation

$$\sigma = \frac{c}{s} = \frac{K}{N_s^{1.447}} \tag{6.24}$$

where K is obtained from Table 6.1.

Table 6.1 Constant for Equation
(6.24) as a Function of Hub-Tip Ratio

D_H/D_t	K
0.3	8.38
0.35	7.65
0.4	6.1
0.45	5.1
0.5	4.49
0.55	3.75
0.6	3.38
0.65	2.94
0.7	2.61

Source: Stepanoff (1957)

The solidity, calculated from (6.24), is based on a suitably chosen value of hub-tip ratio and the required specific speed. It should lie in the range 0.4 to 1.1, and if the calculated solidity lies outside that range, a new choice of D_r/D_t should be made. The optimal number of vanes recommended by Stepanoff(1957) is presented in Table 6.2 as a function of specific speed.

The annular flow area and the required flow rate can now be used to determine the axial velocity component V_a. Thus, we have

$$V_a = \frac{4Q}{\pi(D_t^2 - D_r^2)}$$

(6.25)

Table 6.2 Optimum Number
of Vanes for Axial-Flow Pumps

N_s	n_B
2–3.5	5
3–4.5	4
4–5.5	3
5–6.5	2

Source: Stepanoff (1957)

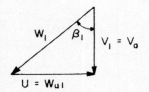

Figure 6.7 Velocity diagram at rotor inlet.

The velocity diagram shown in Figure 6.7 shows the relationship between the mean fluid angle β_1 and the velocities. Because the inlet velocity is axial, it can be determined from (6.25). The mean blade speed U is determined from the mean diameter D_m. Thus, at the mean diameter we can write

$$D_m = \left(\frac{D_t^2 + D_r^2}{2} \right)^{1/2}$$

(6.26)

and

$$U = \frac{ND_m}{2}$$

(6.27)

where N is rotational speed in radians per second. Thus, the required fluid angle β_1 is given by

$$\beta_1 = \tan^{-1} \frac{U}{V_a}$$

(6.28)

Similarly, the fluid angle β_2 at the rotor exit is determined by reference to Figure 6.8. Assuming the same annular flow area, and hence the same axial velocity V_a, we know U and V_a as before. With a known head and with no inlet whirl, i.e., $V_{u1} = 0$, we determine the exit whirl velocity V_{u2} from (6.3); thus, we may write

$$V_{u2} = \frac{gH}{\eta_H U}$$

(6.29)

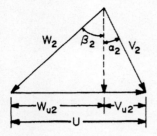

Figure 6.8 Velocity diagram at rotor outlet.

The exit fluid angle β_2 is easily found through the use of the geometric relation

$$\beta_2 = \tan^{-1} \frac{U - V_{u2}}{V_a} \tag{6.30}$$

The mean fluid angle β_m is determined from (6.13), where

$$W_{um} = \frac{W_{u1} + W_{u2}}{2} \tag{6.31}$$

and W_{u1} and W_{u2} are defined by Figures 6.7 and 6.8.

The hydraulic efflciency η_H is equivalent to η_s and can be calculated from equation (6.17). The evaluation of the drag-lift ratio δ from (6.21) requires the use of (6.5) in the form,

$$c_L = \frac{2 \cos \beta_m (\tan \beta_1 - \tan \beta_2)}{\sigma} \tag{6.32}$$

In the absence of cascade data, the profile drag coefficient can be determined from

$$c_D = \frac{\zeta_p \cos^3 \beta_m}{\sigma} \tag{6.33}$$

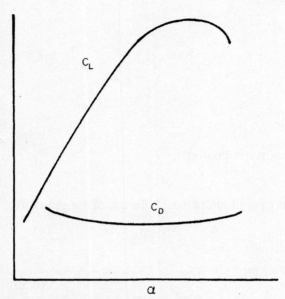

Figure 6.9 Typical cascade results.

Source: Balje (1981)

where the profile loss coefficient ζ_p is extracted from Figure 6.9. The additional drag coefficients are found from (6.18) through (6.20). The stagger angle, i.e., the angle between the chord line of the profile and the axial direction, is determined from the required incidence, or angle of attack α, necessary to produce the lift coefficient c_L calculated from (6.5). Generally, we will select an airfoil section for which cascade data are available. NACA Report 460 is an example of such a source. Figure 6.10 shows schematically the sort of cascade results available in NACA reports and elsewhere. Cascade results should also be checked to assure that the angle of attack chosen does, in fact, produce the desired fluid deflection. Wilson (1984) recommends the use of NASA cascade data for double-circular-arc hydrofoils as a basis for the design of axial-flow pumps and presents carpet plots for this purpose.

The blade may be twisted if the so-called free-vortex method is employed. In this method the product of V_{u2} and D is kept constant, and this variation of V_{u2} with radial position will result in a variation of β_2. The angle β_1 varies with U, and the blade may be twisted to provide proper guidance at the trailing edge, as well as the correct incidence. Free-vortex

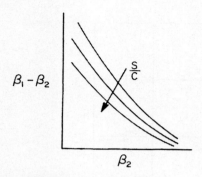

Figure 6.10 Cascade data.

design results in approximately uniform energy transfer at all radial positions. However, untwisted blades may be used in the interest of economy of production. The fluid leaving the rotor blades encounters a row of stationary vanes (Figure 6.11). These serve to straighten the flow, i.e., to remove the whirl component V_{u2} and to increase the pressure. Referring to Figure 6.8, it is seen that fluid enters the vanes at the angle α_2, and leaves axially. The axial component V_a may be reduced by flaring the walls of the annulus by several degrees. The angle formed by the camber line tangent at the leading edge and the vector V_2 should vary from root to tip. It should be designed to provide a positive incidence over the operating range, down to 50 percent of design flow rate.

6.5 Fan Design

The design of an axial-flow fan can proceed in a manner similar to that of the axial-flow pump. Specific speed N_s can be determined from specified values of rotational speed N in rad/s, volume flow rate Q in ft^3/s, and head H in ft. The Cordier curve relation (6.22) may then be used to determine

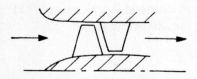

Figure 6.11 Rotor and stator vanes.

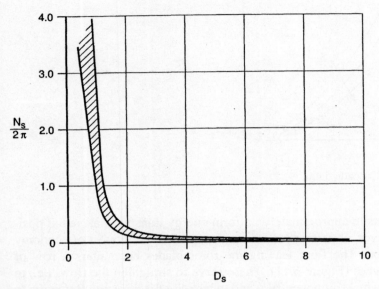

Figure 6.12 Eck's correlation for axial-flow fans.

Source: Eck (1973)

specific diameter D_s. Finally, blade tip diameter D_t is found using (6.23). The root diameter D_r is then calculated from the hub-tip ratio D_r/D_t, which is chosen to lie in the usual range of 0.25 to 0.7. Eck (1973) recommends that the values of specific diameters found in the shaded zone between the curves of Figure 6.12 be used in lieu of those obtained from the Cordier relation for the determination of a tip diameter.

The velocity diagrams, as depicted in Figures 6.7 and 6.8, are then constructed as described in the previous section using (6.25) through (6.31). Fan efficiency, estimated from (6.17), would be expected to lie in the range 0.8–0.9.

The fluid angles β_1 and β_2 required by the velocity diagrams can then be used with cascade data such as those depicted schematically in Figure 6.13 to select a suitable solidity σ for the mean diameter. McKenzie (1988) recommends the following optimal solidity for axial-flow fans:

$$\sigma_{opt} = \frac{1}{9(0.567 - c_{pi})} \qquad (6.34)$$

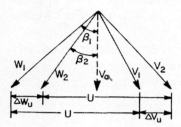

Figure 6.13 Velocity diagram for a compressor stage.

where c_{pi} represents the ideal static pressure rise coefficient and is defined as

$$c_{pi} = 1 - \frac{W_2^2}{W_1^2}$$

(6.35)

for the rotor, and as

$$c_{pi} = 1 - \frac{V_3^2}{V_2^2}$$

(6.36)

for the stator. The actual chord would be selected to provide an aspect ratio h/c of from one to three.

The angle of attack α required to produce c_L calculated from (6.5) is determined from wind tunnel cascade data for the airfoil shape and solidity used in the design. Figure 6.10 schematically shows data of this type, i.e., the blade coefficient as functions of angle of attack α. Angle of attack is the angle between W_1 for a rotor blade or V_2 for a stator blade and the straight line drawn from the leading to the trailing edge of the blade profile; the latter line is also called the chord line. The stagger angle α_s is the angle at which the blade (chord line) is set with respect to the axial direction and is given by $\beta_1 - \alpha$ for rotor blades and by $\alpha_2 - \alpha$ for stator blades.

NACA cascade data have been plotted by Mellor (1956) in a series of charts which enable the determination of β_1 and β_2 for rotor blades, or of α_2 and α_3 for stator blades. There is a separate chart for each solidity and blade profile. Each chart consists of lines of constant stagger angle α_s and lines of constant angle of attack α. The coordinates of the point of intersection of these lines are the desired fluid angles. The use of the Mellor charts is further explained by Wilson (1984).

The blade is twisted to accord with the angles determined from velocity diagrams for the tip and root diameters. Here we use the free-vortex condition, i.e., $V_{u2}D = $ constant, to establish velocity triangles at the extremities of the blade.

6.6   Compressor Design

It is important to design the compressor stage in such a way as to avoid stall. Stall occurs on compressor blades as it does on airplane wings. As the angle of attack of a wing or blade is increased, the lift force increases until a maximum value is achieved; if further increases in angle of attack occur, the wing or blade is said to stall, i.e., to lose lift and, at the same time, to lose pressure rise. The phenomenon of stall occurs as a result of a slowing of the fluid in the boundary layer until the flow stops, or even reverses. Thus, the phenomenon is avoided by maintaining the lift force, or blade loading, below a certain limiting value.

The quantity used as a measure of blade loading and hence of the tendency to stall is the diffusion factor, which, when applied to rotor blades, is defined as

$$D = 1 - \frac{W_2}{W_1} + \frac{\Delta W_u}{2\sigma W_1}$$

$$(6.37)$$

For stator blades W_1 and W_2 are replaced by V_3 and V_2, respectively. Mattingly (1987) recommends the use of designs having $D < 0.55$ to assure avoidance of stall in axial-flow compressors. Wilson (1987) recommends the de Haller criterion, which, for rotor blades, may be stated as $W_2/W_1 < 0.71$, and, for stators, it is $V_3/V_2 < 0.71$.

Equation (6.3) can be expressed in an alternate form as

$$E = UV_u(\tan \beta_1 - \tan \beta_2)$$

$$(6.38)$$

From (6.38) it is clear that if the through flow velocity V_a remains constant, the blade speed U increases with increasing radius, and the energy transfer per unit mass E is to remain independent of radial position, then we must vary the fluid angles β_1 and β_2, and hence, we must vary the blade angles. This is, as previously discussed, the free-vortex condition, $U\Delta V_u =$ constant. As before, with the axial-flow pump, we must have twisted blades in order to achieve this equality of energy transfer along the blade. The variation of blade angles then implies that β_m will vary, and hence the degree of reaction R varies. The latter has been defined by (2.20) and can vary between zero and unity. It is found empirically and has been shown theoretically (Shepherd, 1956) that a value of 0.5 is a near optimum for the degree of reaction producing maximum stage efficiency. Consequently, we find this value frequently used for a design value at the mean diameter. Another design approach is to use a value of 0.5 for R at all radial positions. Both bases for design are used, as well as others not discussed here. It can be shown that $\varphi = \frac{1}{2}$ is also an optimum value of flow coefficient, when the optimum value of R, viz., $R = \frac{1}{2}$, is selected simultaneously.

The theory discussed in the previous section relates energy transfer to fluid angles, blade speed, and axial velocity through the velocity diagrams drawn at the hub, mean, and tip radii. The development of a blade design requires the use of wind tunnel results such as those shown in Figure 6.14 (Herrig et al., 1957). Many such results are available to the designers and they are made for very specific blade shapes. Thus the designer will generally specify the blade shape for which results exist, and these proportions are given in the report of the wind tunnel results. In addition, the tests are carried out for specific values of solidity c/s and stagger angle. For example, Figure 6.14 gives results for the NACA 68 (18): 10 airfoil shape, a solidity 0.75, and a fixed fluid angle β_1 of 60°. For further information on cascade data, the reader is referred to Horlock (1958), Wilson (1984), and Gostelow (1984).

The compressor velocity diagram, of the type shown in Figure 6.15, but in nondimensional form, can be started using the chosen values of R and φ. The mean relative fluid angle β_m can be calculated from (6.13). The blade tip speed can be selected on the basis of the strength of the blade material. Cohen (1987) recommends the stress equation,

$$s_c = 0.5\rho_B U_t^2 \left(1 - \frac{D_H^2}{D_t^2}\right)$$

$$(6.39)$$

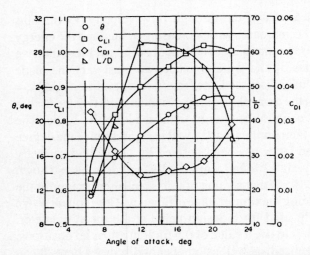

Figure 6.14 Cascade results for the NACA Profile 68 (18): 10.

Source: Herrig et al. (1957).

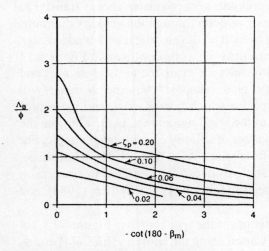

Figure 6.15 Profile loss coefficients.

Source: Balje (1981).

where s_c is the design centrifugal stress, and ρ_B is the density of the blade material. To solve for the tip speed, a tentative value of hub-tip ratio must be chosen. Wilson (1984) recommends hub-tip ratios of greater than 0.6 for axial-flow compressors.

Substituting the chosen hub-tip ratio and a design centrifugal stress for the blade material into (6.39), a safe tip speed and tip radius can be determined. Since a hub-tip ratio was also used, both r_t and r_H can be determined. Tip speeds of 1500 fps and less are typical. Finally, the mean blade speed U is determined from the expression,

$$U = \frac{r_m U_t}{r_t} \tag{6.40}$$

where r_m is determined from (6.26).

The mean velocity is used to convert the dimensionless velocities into dimensional ones. The blade loading ΔW_u, or ΔV_u is obtained from the energy transfer using (6.3). Since the energy transfer is required, it must be obtained from the required stage total pressure ratio R_{pn} through the use of the definition of the stage efficiency, viz.,

$$R_{pn} = \left(1 + \frac{\eta_s \Delta T_{os}}{T_{o1}}\right)^{\gamma/(\gamma - 1)} \tag{6.41}$$

and the steady flow energy equation,

$$E = c_p \Delta T_{os} \tag{6.42}$$

where ΔT_{os} represents total temperature rise for the stage. A value of stage efficiency must be assumed at this point, and this requires later verification of the assumed value. The above calculations result in the determination of all the velocities and angles in the velocity diagram.

At this point several checks must be made. The diffusion factor should be calculated to assure that $D < 0.55$. A solidity is required in (6.37); one should be selected in the range of 0.66 to 2.0. A typical value of solidity is 1.0. The hub-tip ratio must be checked to assure that the mass flow rate is the required value. Equation (6.25) can be used for this purpose along with the volume flow relation, $Q = \dot{m}/\rho$. Finally, the assumed stage efficiency is

Table 6.3 Optimum Camber Angles in Degrees for Rotor and Stator Blades of Axial Flow Compressors

Deflection, ε degrees	s/c		
	0.5	1.0	1.5
15	–	14	22
20	15	27	33
25	27	38	46
30	35	47	–

Source: Horlock (1958)

checked using (6.17). For this calculation, the aspect ratio can be selected in the range of $1 < h/c < 3$, and the tip clearance ratio can be taken as 0.02.

The choice of blade can be made on the basis of the camber angles presented in Table 6.3. The camber angle is the difference between the angles formed by tangents to the camber line at the ends of the blade profile (Figure 6.16). Thus,

$$\theta = \gamma_1 - \gamma_2 \tag{6.43}$$

Once the blade having the selected camber angle is selected, the deviation, defined by

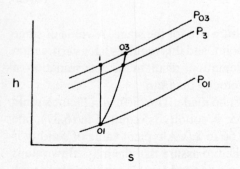

Figure 6.16 Enthalpy–entropy diagram.

$$\delta' = \beta_2 - \gamma_2 \tag{6.44}$$

can be calculated from the Carter rule, viz.,

$$\delta' = \frac{\theta}{4\sigma^{1/2}} \tag{6.45}$$

which is recommended by Mattingly (1987). The angle γ_2 is then calculated from (6.44), and γ_1 is determined from (6.43). Finally, the stagger angle is determined from the approximate relation,

$$\alpha_s = \frac{\gamma_1 + \gamma_2}{2} \tag{6.46}$$

The recommended calculations relate to rotor design; however, for stator blade design, the same procedure is followed, but V_2, V_3, α_2, and α_3 replace W_1, W_2, β_1, and β_2, respectively.

6.7   Compressor Performance

Prior to construction and testing of the prototype machine it is desirable to determine estimated performance characteristics by means of calculation. Normally, stage efficiency η_s as well as multistage compressor efficiency are based on total temperatures. Thus, referring to Figure 6.17, we have

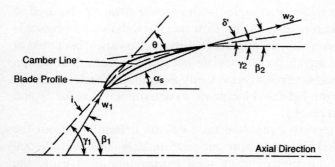

Figure 6.17 Compressor blade diagram.

$$\eta_s = \frac{T_i - T_{o1}}{T_{o3} - T_{o1}} \tag{6.47}$$

This is the same definition used in (5.4) for centrifugal compressors. In Figure 6.17, state 1 denotes conditions at the rotor inlet and state 3 those at the stator outlet. It has been shown, however, by Cohen et al. (1974) that the incompressible definition (6.17) predicts the stage efficiency well, because the rise of total temperature in the stage is sufficiently small. Cascade test results may be used to determine values of lift and drag coefficients for the blade profile. If cascade data are not available, the lift and drag coefficients can be calculated from (6.18–6.20), (6.32), and (6.33).

The energy transfer relation (6.3) is quite accurate for a single stage, because the velocity profile $V_a(r)$ is nearly flat. The annular walls create a boundary layer which causes peaking of V_a near the mean radius. Thus (6.3) must be multiplied by a factor λ, called the work-done factor, and the resulting equation,

$$E = \lambda U V_a(\tan \beta_1 - \tan \beta_2) \tag{6.48}$$

can be used for each stage with a constant value of λ. The work-done factor may be approximated from

$$\lambda = 0.85 + 0.15 \exp\left[\frac{-(N_{st} - 1)}{2.73}\right] \tag{6.49}$$

where N_{st} is the number of stages in the compressor.

The same basic relations, coupled with cascade data, can be used to predict off-design performance. As the flow rate through the compressor is varied from the design value, the angle of incidence also varies, but the rotor and stator fluid exit angles do not deviate appreciably from their design values. Thus it is possible to construct velocity diagrams for each off-design flow rate, and from the indicated incidence to determine values of c_L and c_D from cascade test results.

The overall compressor pressure ratio can be determined from the product of individual stage pressure ratios. Similarly, the overall total temperature rise is the sum of stage total temperature rises using the definitions of (6.41) and (6.42) applied to the compressor as a whole. Thus,

$$\eta_c = \frac{R_p^{(\gamma-1)/\gamma} - 1}{\Delta T_o / T_{in}}$$

(6.50)

where R_p and ΔT_o in this equation denote the total pressure ratio and the total temperature rise for the whole machine. The polytropic efficiency η_p is sometimes used in multistage compressors to calculate the outlet temperature; thus, $(\gamma - 1)/\eta_p\gamma$ replaces the polytropic exponent $(n - 1)/n$, so that

$$T_{out} = T_{in} R_p^{(\gamma-1)/\eta_p \gamma}$$

(6.51)

The compressor map for an axial-flow compressor will have the same appearance as that shown schematically for the centrifugal compressor in Figure 5.5. This plot of R_p as a function of $\dot{m}$ with N as parameter shows operational limits set by the phenomena of stalling at low flow rates and choking at high flow rates. At low speeds, choking occurs in the rear stages and stalling (due to high incidence) in the front stages, whereas the situation is reversed at high rotor speeds. These phenomena can be predicted in advance using indicators such as the critical Mach number M_c based on inlet relative velocity (usually $M_c \approx 0.7$–0.8) to indicate the first appearance of sonic flow in the blade passages, and the stalling incidence angle corresponding to the maximum value of c_L obtained in cascade tests. Since temperatures increase in stages after the first, Mach numbers decrease. Thus the first stage will be the most likely site of shock losses. The first stage may be designed for supersonic inlet velocities near the tips. The leading edge of such blades will be sharp to accommodate attached oblique shocks as discussed by Kerrebrock (1977). The blades are called transonic in that they accommodate subsonic flow near the hub. Such a stage may be desirable in aircraft compressors where the cross-sectional area is minimal.

6.8 Examples

Example Problem 6.1

In an axial-flow compressor stage the degree of reaction is 0.5, the mean blade speed is 1000 fps, the flow coefficient is 0.4 and the blade loading coefficient is 0.355. Lift and drag coefficients are 1.30 and 0.055, respec-

tively, where c_d includes all losses. The entering air density is 0.0024 slug/cu ft. Determine the energy transfer and the stage static pressure rise.
Solution: Calculate the change in the whirl component of velocity.

$$\Delta V_u = \Lambda_B U = (0.355)(1000) = 355 \text{ fps}$$

Calculate the energy transfer.

$$E = \Delta V_u U = (355)(1000) = 355,000 \text{ ft}^2/\text{s}^2$$

Calculate drag-lift ratio.

$$\delta = \frac{c_d}{c_L} = \frac{0.055}{1.30} = 0.0423$$

Calculate the stage pressure rise.

$$P_3 - P_1 = \rho U^2 \Lambda_B \varphi \left[\frac{R - \varphi\delta}{\varphi + \delta R} + \frac{1 - R - \varphi\delta}{\varphi + \delta(1 - R)} \right]$$

$$= (0.0024)(1000)^2(0.355)(0.4) \left[\frac{0.5 - (0.4)(0.0423)}{0.4 + (0.0423)(0.5)} + \right.$$

$$\left. \frac{1 - 0.5 - (0.4)(0.0423)}{0.4 + (0.0423)(1 - 0.5)} \right] = 783 \text{ psf} = 5.43 \text{ psi}$$

Example Problem 6.2

An axial-flow, high-performance, single-stage, experimental air compressor [see Paulon et al. (1991)] has no inlet guide vanes and runs at 12,000 rpm. The tip blade speed is 1322 fps, the hub-tip ratio is 0.70, and the solidity is 1.375. Find the mean blade speed.
Solution: Calculate the tip radius.

$$r_t = \frac{U_t}{N} = \frac{1322}{12,000(\pi/30)} = 1.052 \text{ ft}$$

The circle of mean radius divides the annular area of the compressor into two equal areas. Thus, the mean radius is calculated from the following equation:

$$r_m = \left(\frac{r_t^2 + r_H^2}{2} \right)^{1/2} = \left[\frac{1.49(1.052)^2}{2} \right]^{1/2} = 0.908 \text{ ft}$$

where $r_H = 0.7r_t$ has been used. Finally, the mean blade speed is

$$U = Nr_m = (12000)(\pi/30)(0.908) = 1141$$

Example Problem 6.3

Using the data from Example Problem 6.2, find the absolute and relative Mach numbers of the air entering the rotor, if the relative air angle at entry is 60° and T_{o1} is 540°R.
Solution: Noting that V_1 is axially directed, the velocity diagram is as shown in Figure 6.7. Thus,

$$V_1 = U \cot \beta_1 = 1141 \cot 60° = 659 \text{ fps}$$

$$W_1 = U \csc \beta_1 = 1141 \csc 60° = 1317.5 \text{ fps}$$

Calculate the static temperature using the relation between static and total temperature.

$$T_1 = T_{o1} - \frac{V_1^2}{2c_p} = 540 - \frac{(659)^2}{12012} = 504°R$$

Calculate the acoustic speed in the inlet air.

$$a_1 = (\gamma R T_1)^{1/2} = [(1.4)(1716)(504)]^{1/2} = 1100 \text{ fps}$$

Finally, calculate the Mach numbers.

$$M_1 = \frac{V_1}{a_1} = \frac{659}{1100} = 0.60$$

$$M_{R1} = \frac{W_1}{a_1} = \frac{1317.5}{1100} = 1.198$$

Example Problem 6.4

The relative air angle at the rotor exit is 35° in the compressor stage considered in Example Problems 6.2 and 6.3. Find the energy transfer and stage total pressure ratio, assuming a stage efficiency 0.87. Note: an efficiency this high is realizable in modern, high-performance stages.
Solution: A previous calculation gave $V_1 = 659$ fps (axially directed). From Figure 6.7, it is clear that

$$W_{u1} = U = 1141 \text{ fps}$$

Referring to Figure 6.8, it is seen that

$$W_{u2} = V_1 \tan \beta_2 = (659) \tan 35° = 461 \text{ fps}$$

The energy transfer is given by (6.3) and is

$$E = U\Delta V_u = 1141(1141 - 461) = 775,880 \text{ ft}^2/\text{s}^2$$

The stage pressure ratio is given by (6.41) and (6.42) and is

$$R_p = \frac{p_{o3}}{p_{o1}} = \left(1 + \frac{\eta_s E}{c_p T_{o1}}\right)^{\gamma/(\gamma-1)}$$

$$R_p = \left[1 + \frac{(0.87)(775880)}{(6006)(540)}\right]^{3.5} = 1.938$$

Note: Pressure ratio measured by Paulon et al. (1990) for the same design was 1.95.

Example Problem 6.5

Estimate the stage efficiency for the single-stage compressor described in Example Problems 6.2–6.4. The approximate rotor solidity is 1.375, the approximate rotor blade aspect ratio is 0.72 and the assumed tip clearance ratio c_T/h is 0.02.

Solution: Calculate degree of reaction.

$$-W_{um} = \frac{-W_{u1} - W_{u2}}{2} = \frac{1141 + 461}{2} = 801 \text{ fps}$$

$$R = \frac{-W_{um}}{U} = \frac{801}{1141} = 0.702$$

Calculate the flow coefficient.

$$\varphi = \frac{V_1}{U} = \frac{659}{1141} = 0.5776$$

Calculate the mean relative air angle.

$$\tan \beta_m = \frac{\tan \beta_1 + \tan \beta_2}{2} = \frac{\tan 60° + \tan 35°}{2}$$

$$\beta_m = 50.57°$$

Solve for c_L using equation (6.32).

$$c_L = 2 \cos \beta_m \frac{\tan \beta_1 - \tan \beta_2}{\sigma}$$

$$c_L = 2 \cos 50.57° \frac{\tan 60° - \tan 35°}{1.375} = 0.953$$

Use Figure 6.15 to determine the profile loss coefficient. First, find blade loading coefficient over flow coefficient.

$$\frac{\Lambda_B}{\varphi} = \frac{\Delta V_u}{V_a} = \frac{W_{u1} - W_{u2}}{V_a} = \frac{1141 - 461}{659} = 1.03$$

Next, find the chart abscissa.

$$- \cot (180 - \beta_m) = - \cot (180 - 50.57) = 0.82$$

Enter the chart and find $\zeta_p = 0.09$. Calculate c_D for profile drag from equation (6.33).

$$c_D = \frac{\zeta_p}{\sigma} \cos^3 \beta_m = \frac{0.09}{1.375} \cos^3 (50.57) = 0.0168$$

Next, apply (6.18), (6.19), and (6.20) to determine additional contributions to drag.

$$c_{D'} = 0.02 \frac{s}{h} = \frac{0.02}{\sigma h/c} = \frac{0.02}{(1.375)(0.72)} = 0.20$$

$$c_{D''} = 0.018 c_L^2 = 0.018(0.953)^2 = 0.01635$$

$$c_{D'''} = 0.29 \left(\frac{c_T}{h} \right) c_L^{3/2} = 0.29(0.02)(0.953)^{3/2} = 0.0054$$

Next, solve for the total drag over the lift using (6.21).

$$\delta = \frac{0.0168 + 0.020 + 0.01635 + 0.0054}{0.953} = 0.062$$

Finally, compute the stage efficiency using (6.17).

$$\eta_s = 0.5776 \left[\frac{0.702 - (0.5776)(0.062)}{0.5776 + (0.062)(0.702)} + \frac{1 - 0.702 - (0.062)(0.5776)}{0.5776 + (0.062)(0.298)} \right] = 0.873$$

Note: It is interesting that the stage efficiency measured by Paulon et al. (1990) for the corresponding actual machine was 0.872.

Example Problem 6.6

Find the overall compressor efficiency for a four-stage compressor, in which each stage has the same velocity diagram and stage efficiency. The inlet temperature of the compressor is 540°R and the velocity diagrams are the same as in Example Problems 6.2 through 6.4.

Solution: Since we have a multistage machine, we must apply a work done factor. Thus, from (6.48) and (6.49) with $N_{st} = 4$,

$$\lambda = 0.85 + 0.15 \exp\,[-(N_{st} - 1)/2.73] = 0.90$$

and

$$E = (0.90)(1141)(1141 - 461) = 698{,}292 \text{ ft}^2/\text{s}^2$$

The individual stage pressure ratios are calculated from equations (6.41) and (6.42) using the above E and the given η_s. Thus,

$$R_{pn} = \left[1 + \frac{(698{,}292)(0.87)}{(6006)T_{on}}\right]^{3.5}$$

applies to each of the four stages, when n takes on the value of 1, 2, 3, or 4, corresponding to the stage number. The rise of total temperature is the same across any stage and is

$$\Delta T_{os} = \frac{E}{c_p} = \frac{698{,}292}{6006} = 116.27°R$$

The overall temperature rise for four stages is

$$\Delta T_o = 4\Delta T_{os} = 4(116.27) = 465.08°R$$

The inlet total temperature T_{on} increases by 116.27° in each subsequent stage. Thus, the inlet temperatures for the four stages are the following: $T_{o1} = 540°R$, $T_{o2} = 656.27°R$, $T_{o3} = 772.54°R$, and $T_{o4} = 888.81°R$.

Substituting these values into the pressure-ratio equation yields the following total pressure ratios for the four stages: $R_{p1} = 1.824$, $R_{p2} = 1.652$,

$R_{p3} = 1.538$, and $R_{p4} = 1.458$. The product of these pressure ratios is the overall pressure ratio for the compressor, which is

$$R_p = (1.824)(1.652)(1.538)(1.458) = 6.757$$

The overall pressure ratio and the overall temperature rise are used in equation (6.50) to determine the overall efficiency of the four-stage compressor. Substituting, we have

$$\eta_c = \frac{(6.757)^{1/3.5} - 1}{(465.08/540)} = 0.843$$

It is noted that the overall efficiency is a little lower than the stage efficiency, which is always the case for multistage compressors, and is just the opposite for multistage turbines. For multistage machines with small stage pressure ratios, it is sometimes assumed that the stage efficiency is equal to the polytropic efficiency, when the number of stages is five or more. In the present example the stage pressure ratios are unusually high, and the number of stages is only four; thus, the approximation would not be expected to apply. However, we will calculate the polytropic efficiency to make a comparison with the stage efficiency for the present case. First, we calculate the compressor outlet temperature in the following way:

$$T_{out} = T_{o1} + \Delta T_o = 540 + 465.08 = 1005°R$$

Then, the polytropic efficiency is calculated from equation (6.51) and is

$$\eta_p = \frac{(\gamma - 1)\ln R_p}{\gamma \ln (T_{out}/T_{o1})} = \frac{\ln (6.757)}{(3.5) \ln (1005/540)} = 0.8788$$

Note: Even in the present case the polytropic and stage efficiencies are nearly equal.

References

Balje, O. E. 1981. *Turbomachines*. John Wiley & Sons, New York.

Cohen, H., G. F. C. Rogers, and H. I. H. Saravanamuttoo. 1987. *Gas Turbine Theory*. Longman, London.

Eck, B. 1973. *Fans*. Pergamon, Oxford.

Gostelow, J. P. 1984. *Cascade Aerodynamics*. Pergamon, Oxford.

Herrig, J., J. C. Emery, and J. R. Erwin. 1957. *Systematic Two-Dimensional Cascade Tests of NACA 65-Series Compressor Blades at Low Speeds*. NACA Tech. Note No. 3916, National Aeronautics and Space Administration, Washington, DC

Horlock, J. H. 1958. *Axial Flow Compressors*. Butterworths, London.

Kerrebrock, J. L., 1977. *Aircraft Engines and Gas Turbines*, MIT Press, Cambridge, Mass.

Mattingly, J. D. 1987. *Aircraft Engine Design*. AIAA, New York.

McKenzie, A. B. 1988. The selection of fan blade geometry for optimum efficiency. *Proceedings of the Institution of Mechanical Engineers*. IME, United Kingdom.

Mellor, G. 1956. *The NACA 65-Series Cascade Data*. MIT Press, Cambridge, Mass.

Paulon, J., et al. 1991. *Design and Test Results of a High Performance Single Stage Compressor*. Office National D'Etudes et de Recherches Aerospatiales No. 1991-176. ONERA, Chatillon.

Shepherd, D. G. 1956. *Principles of Turbomachinery*, Macmillan, New York.

Stepanoff, A. J. 1957. *Centrifugal and Axial Flow Pumps*. John Wiley & Sons, New York.

Wilson, D. G. 1984. *The Design of High-Efficiency Turbomachinery and Gas Turbines*. MIT Press, Cambridge.

Wislicenus, G. F. 1965. *Fluid Mechanics of Turbomachinery*. Dover, New York.

Problems

6.1   An axial-flow fan is to be designed with a tip diameter of 9.5 in. and a hub-tip ratio of 0.5. Assuming that the fan is driven by a 1.5 hp motor and has an overall efficiency of 80 percent, determine the flow rate and desirable speed. The fan discharges air into the room through an exit area of 78.54 in.^{2}.

6.2   Construct velocity diagrams for the rotor inlet and outlet at the mean diameter for the fan considered in Problem 6.1.

6.3   Find the degree of reaction at the hub and tip of an axial-flow compressor stage of free-vortex design, hub-tip ratio of ⅓ and flow coefficient at the hub of 1.0. In and out of the stage the flow is purely axial. At the hub the flow is turned 30° in the rotor blades. The flow is modeled as incompressible.

6.4   Given the data of Problem 6.3, show that the degree of reaction increases from hub to tip and can be written as

$$R = 1 - \frac{E}{2(Nr)^2}$$

6.5   Find the degree of reaction at the tip of an axial-flow compressor stage of free-vortex design, hub-tip ratio 0.357, flow coefficient at the hub of 1.25, hub solidity of 1.89, hub degree of reaction of −0.649 and $E = V_a^2$. The blade chord is constant from hub to tip. The flow is modeled as incompressible. Assume that $V_1 = V_3$.

6.6   Given the data of Problem 6.5, determine the relative and absolute flow angles at tip and hub. Draw the velocity diagrams for tip and hub.

6.7   Given the data of Problem 6.5, find the diffusion factor at the tip of the rotor and at the hub of the stator.

6.8   Find the diffusion factor at hub and tip for both rotor and stator of an axial-flow compressor stage of free-vortex design, hub-tip ratio of 0.385, flow coefficient at the hub of 1.43, hub solidity of 1.695, hub degree of reaction of −0.599 and $E = V_a^2$. Assume that $V_1 = V_3$. The flow is modeled as incompressible.

6.9   Given the data of Problem 6.8, determine the relative and absolute flow angles at tip and hub. Draw the velocity diagrams for tip and hub.

6.10   In an axial-flow compressor stage R = 0.5, U = 1030 fps, V_a = 400 fps, W_1 = 800 fps, and W_2 = 523 fps. Lift and drag coefficients are the same in rotor and stator and are 1.4 and 0.08, respectively. c_d includes all losses. Calculate stage pressure rise, if the entering gas density is 0.0040 slugs/cu ft.

6.11   In the first stage of an axial-flow air compressor mean values are U = 181 m/s, V_a = 151 m/s, and R = 0.5. Air in the room at 1.01 bar and 287°K is drawn into the compressor at the rate of 19.98 kg/s. Rotational speed is 9000 rpm. The total temperature leaving the stage 308°K. The work-done factor for the stage is 0.961. Find:
        a) the relative air angles
        b) the mean radius
        c) the blade length

6.12 Use the cascade results of Figure 6.14 to determine the velocity triangles (mean radius) and static pressure rise for a compressor stage having the following features: $U = 1000$ ft/s, $\Delta T_{oS} = 54°F$, $\beta_1 = 60°$, $\rho = 0.00237$ slug/ft^3, $\alpha = 12°$; $\sigma = 0.75$, and h/c = 2.

6.13 Determine the mean radius, air angles, and blade length for the first stage of a compressor having the following data: $N = 150$ rev/s; $\Delta T_{oS} = 36°F$, $\dot{m} = 44$ lb/s, $V_a = 492$ ft/s, $U = 590$ ft/s, $\lambda = 0.96$, $R = 0.5$ (mean radius), $P_{o1} = 1$ atm, $T_{o1} = 518°R$.

6.14 A axial-flow air compressor stage with a solidity of 0.75, an aspect ratio of two, an inlet relative air angle $\beta_1 = 60°$, an outlet relative air angle $\beta_2 = 42°$, a flow coefficient of 0.4, inlet air density of 0.00237 slugs/cu ft, and a mean blade speed of 1050 fps. Use the velocity diagram at the mean radius to find:

   a) c_L
   b) R
   c) D
   d) θ/c
   e) ζ_p
   f) $c_{D'}$
   g) $c_{D''}$
   h) $c_{D'''}$
   i) η_s

6.15 Use the cascade results for the NACA 68(18):10 airfoil with a solidity of 0.75, inlet relative gas angle of 60° and an outlet relative gas angle of 42°. The mean blade speed is 1000 fps, $T_{o1} = 520°R$, the stage total temperature rise is 54°R and $P_{o1} = 14.1$ psia. Take the gas to be air with a specific heat $c_p = 5999$ ft-lb/slug-°R. Take $V_1 = V_3$ and V_a as a constant. Take the aspect ratio to be two. The ratio of tip clearance to blade height is 0.02. Find:

   a) E
   b) V_a
   c) V_1
   d) V_2
   e) the air density entering the rotor
   f) the flow coefficient at the mean radius
   g) the blade loading coefficient at mean radius
   h) R at mean radius
   i) D/L (delta) based on the four losses
   j) pressure rise for the stage

   k) density at stage exit

   l) stage efficiency

   m) stage total pressure ratio from incompressible model

   n) stage total pressure ratio from compressible model

6.16 Repeat Problem 6.15 without the use of cascade data.

6.17 Air at $P_{o1} = 14.7$ psia and $T_{o1} = 519°R$ enters a three-stage compressor with a velocity of 350 ft/s. There are no inlet guide vanes, and the axial component V_a remains constant through the compressor. In each stage the rotor turning angle is 25 degrees. The annular flow passages are shaped so that the mean blade radius is everywhere 9 inches. The rotor speed is 9000 rpm. The ratio of specific heats is 1.4, and the polytropic efficiency is constant at 0.90. The blade height at the inlet is 2 inches. Draw the velocity diagram below and calculate:

   a) the work per unit mass for each stage in ft-lb/slug

   b) the mass flow rate of air in lb/s

   c) the power to run the compressor

   d) stage temperature ratios

   e) the overall compressor pressure ratio

   f) the blade height at the exit from the third stage

6.18 Air enters an axial-flow compressor stage axially (no inlet guide vanes) at $P_{o1} = 14.696$ psia and $T_{o1} = 60°F$ with $V_1 = 490$ fps. The air is turned $30°$ at the mean radius by the rotor blades, and the stator blades turns the air to the axial direction, i.e., $V_1 = V_3 = V_a$. The tip of the rotor blade has a radius of 12 inches and the blade length is 2 inches. Rotational speed is 6000 rpm. The stage efficiency is 0.90. Find:

   a) mean radius of blade

   b) mean blade speed

   c) relative air angles

   d) absolute air angles

   e) energy transfer

   f) mass flow rate of air

   g) power required to drive the stage

6.19 The first stage of an axial-flow compressor receives air from the atmosphere, which is at 14.7 psia and 520°R when at rest. At the mean radius the inlet guide vanes turn the air 30 degrees from the axial direction; moreover, the moving vanes reduce the relative air angle by 30 degrees, while the mean blade speed is 1000 fps. The degree of reaction is 0.5, and the stage efficiency is 0.90. The mean radius is 10 inches, and the blade length is 2 inches. Neglect the total pressure loss in the inlet guide vanes. Find:

    a) the rotational speed
    b) the relative air angles
    c) the flow coefficient
    d) the energy transfer
    e) the mass flow rate
    f) the power required to operate the stage
    g) the total-to-total pressure ratio for the stage

6.20 Each stage of a three-stage, axial-flow air compressor has the same velocity diagram, same stage efficiency, same mean radius, and the same rotational speed as in Problem 6.19. The same first-stage dimensions and inlet air conditions are also applicable. Neglect the total pressure loss in the inlet guide vanes. Determine:

    a) the overall ratio of total pressure
    b) the power to drive the compressor
    c) the total temperature at the compressor discharge

6.21 An axial-flow compressor is to be designed to handle 6.6 lb/s of air, which is drawn from an ambient state of 14.7 psia and 520°R. The overall ratio of total pressure is to be four, and the total temperature rise per stage is to be 41.5°R, while the rotational speed is 20,000 rpm. Choose a constant mean radius of 4.5 inches. Assume a polytropic efficiency of 0.88. Use optimum values of flow coefficient and degree of reaction, viz., $\varphi = 0.5$ and $R = 0.5$ for all stages. Determine:

    a) number of stages required
    b) length of first-stage rotor blades
    c) length of last-stage rotor blades

6.22 Derive equation (6.5).

6.23 Show that the optimum degree of reaction for an axial-flow compressor stage is $\frac{1}{2}$. Hint: Find the minimum value of the ratio of rate of loss of mechanical energy, $\dot{E}_L = W_m D_R + V_m D_S$, to rate of energy input, given by $\dot{E} = U(L_R \cos \beta_m + D_R \sin \beta_m)$. Assume that $D_R \sin \beta_m \ll L_R \cos \beta_m$, $L_R \cos \beta_m = L_S \cos \alpha_m$, and $\delta_R = \delta_S$.

6.24 Show that when $R = \frac{1}{2}$, i.e., when the degree of reaction is optimal, the mean values of the relative and absolute gas angles are equal, that $V_m = W_m$ and that $W_{um} = -V_{um}$

6.25 Using the ratio $\dot{E}_L/\dot{E}$, show that, when $R = \frac{1}{2}$, the optimum value of flow coefficient is also $\frac{1}{2}$.

6.26 Show that the profile loss coefficient ζ_p is related to the blade drag coefficient through equation (6.33). Hint: Use the control volume of Figure 6.5, equation (6.9), equation (6.15), and the force relationship, $D = F_{Bu} \sin \beta_m - F_{Ba} \cos \beta_m$. Define c_D as $2D/\rho W_m^2 c$.

6.27 Derive an equation which is similar to (6.11), but expresses the pressure rise in the stator. Assume $V_3 = V_1$. Hint: Note that α_m replaces β_m in the stator analysis. Moreover, the lift force on the stator blade is perpendicular to the mean velocity V_m.

6.28 Derive equation (6.15).

6.29 Derive equation (6.14).

6.30 Derive equation (6.17).

6.31 Show that, when no frictional processes are present, (6.17) reduces to
$$\Delta p_{ideal} = \rho U^2 \Lambda_B$$

6.32 The following data are available for an axial-flow water pump: $\varphi = 0.333$, $\beta_1 = 72°$, $\beta_2 = 55°$, $\sigma = 1.47$, $h/c = 1.0$. Estimate the hydraulic efficiency. Hint: Use equation (6.17).

6.33 An axial-flow fan is to be designed to deliver 26,500 cfm (cu ft/min) while the rotor turns at 1500 rpm. The rise in total pressure across the fan is to be 1.18 inches of water, and the air density is 0.076 lb_m/cu ft. Determine the hub and tip diameters. Hint: Use Figure 6.12 and choose a hub-tip ratio of 0.3.

6.34 Derive the free-vortex condition, $V_u r =$ constant, which is used in the design of axial-flow machines and is discussed in Section 6.6. Assume that, in the flow field, V_a is independent of axial and radial position, the flow is frictionless, $V_r = 0$ everywhere and the pressure forces balance the centrifugal forces at all points.

Symbols for Chapter 6

c	chord length = distance from leading edge to trailing edge of blade profile
c/s	solidity = σ
c_D	profile drag coefficient
$c_{D'}$	annulus drag coefficient
$c_{D''}$	secondary flow drag coefficient
$c_{D'''}$	tip clearance drag coefficient
c_L	lift coefficient
c_T	tip clearance

c_p specific heat of gas at constant pressure

c_{pi} ideal static pressure rise coefficient

c_v specific heat of gas at constant volume

D diffusion factor

D_H hub diameter

D_m diameter at mean position between hub and tip of blade

D_R drag force on rotor blade = component of F_B in direction of W_m

D_r root diameter = D_H

D_S drag force on stator blade = component of F_B in direction of V_m

D_s specific diameter

D_t tip diameter

E energy transfer = $U\Delta V_u = U\Delta W_u$

$\dot{E}$ rate of energy transfer

$\dot{E}_L$ rate of loss of fluid mechanical energy to thermal energy

F_B blade force per unit length of blade

F_{Ba} axial component of F_B

F_{Bu} tangential component of F_B

h blade height = $r_t - r_H$

h/c aspect ratio

H output head = $\eta_H U(V_{u2} - V_{u1})/g$

i incidence = $\beta_1 - \gamma_1$

L_R lift force on rotor blade = component of F_B perpendicular to W_m

L_S lift force on stator blade = component of F_B perpendicular to V_m

h_o specific total (stagnation) enthalpy of the fluid

$\dot{m}$ mass flow rate of fluid between adjacent blades per unit blade length

N rotor speed

N_s specific speed

n polytropic exponent ($n = \gamma$ for isentropic processes)

p fluid pressure

p_{o1} total pressure at rotor inlet

p_{o2} total pressure at rotor exit or at stator inlet

p_{o3} total pressure at stator exit

R degree of reaction

r_H hub radius or radius at root of blade

r_m mean blade radius = $D_m/2 = (r_t^2 + r_H^2)^{1/2}$

r_t radius at blade tip

R_p overall total pressure ratio of a multistage compressor

R_{pn} total pressure ratio of a single compressor stage

S	blade width
s	spacing between adjacent blades = pitch
s_c	design centrifugal stress in blade
s/c	pitch-chord ratio
T	absolute temperature
T_{o1}	total temperature at rotor inlet
T_{o2}	total temperature at rotor outlet or at stator inlet
T_{o3}	total temperature at stator outlet
T_i	total temperature at the end of an isentropic compression from p_{o1} to p_{o3}
T_{in}	total temperature at inlet to first stage of a multistage compressor
T_{out}	total temperature at outlet of last stage of a multistage compressor
ΔT_o	overall rise of total temperature in multistage compressor
ΔT_{os}	total temperature rise in a single stage of a compressor
U	blade speed at mean radius r_m of blade
U_t	blade speed at tip radius r_t of blade
U_H	blade speed at hub radius r_H of blade
V	absolute velocity of fluid
V_m	mean absolute velocity $= (V_{um}^2 + V_a^2)^{1/2}$
V_1	absolute velocity of fluid entering rotor
V_2	absolute velocity of fluid leaving rotor or entering stator
V_3	absolute velocity of fluid leaving stator
V_a	component of V in axial direction
V_r	component of V in radial direction
V_u	component of V in tangential direction
V_{u1}	tangential component of V_1
V_{u2}	tangential component of V_2
ΔV_u	$V_{u2} - V_{u1}$
V_{um}	tangential component of $V_m = (V_{u1} + V_{u2})/2$
W	velocity of fluid relative to moving blade
W_1	relative velocity entering rotor
W_2	relative velocity leaving rotor
W_m	mean relative veloclty $= (V_a^2 + W_{um}^2)^{1/2}$
W_u	tangential component of W
ΔW_u	$W_{u1} - W_{u2} = \Delta V_u$
W_{u1}	tangential component of W_1
W_{u2}	tangential component of W_2
W_{um}	tangential component of $W_m = (W_{u1} + W_{u2})/2$

α angle of attack = angle between W_1 and chord line for moving blade

α_s stagger angle

α_1 angle between V_1 and V_a

α_2 angle between V_2 and V_a

α_3 angle between V_3 and V_a

α_m angle between V_m and $V_a = \tan^{-1}(V_{um}/V_a)$

β_1 angle between W_1 and V_a

β_2 angle between W_2 and V_a

β_m angle between W_m and $V_a = \tan^{-1}(W_{um}/V_a)$

γ ratio of specific heats = c_p/c_v

γ_1 angle between leading edge tangent to camber line of a blade profile and the axial direction

γ_2 angle between trailing edge tangent to camber line of a blade profile and the axial direction

δ drag-lift ratio

δ' deviation = $\beta_2 - \gamma_2$

ϵ fluid deflection by blade = $\beta_1 - \beta_2$ or $\alpha_2 - \alpha_3$

η_c compressor efficiency

η_H hydraulic efficiency [see (4.15)]

η_p polytropic efficiency

η_s stage efficiency

θ camber angle = $\gamma_1 - \gamma_2$

Λ_B blade loading coefficient

λ work-done factor

ρ fluid density

ρ_B density of solid blade material

σ solidity at mean radius of blade = c/s

σ_H solidity at hub = solidity at root of blade

σ_t solidity at tip of blade

φ flow coefficient = V_a/U

7   Radial-Flow Gas Turbines

7.1   Introduction

The simplest gas turbine engine requires at least two major components besides the turbine proper (Figure 7.1). The gas (usually air) must be compressed by a centrifugal or axial-flow compressor, and then it must be heated (usually by burning a hydrocarbon fuel) in a combustor or heat exchanger. Gas is delivered to the turbine inlet at an elevated pressure and temperature.

The ideal thermodynamic cycle associated with the simple gas turbine is the Brayton cycle depicted in Figure 7.2. Process 1–2 is an isentropic compression, 2–3 an isobaric heating, and 3–4 an isentropic expansion. A more realistic model of the gas processes would follow the dashed lines 1–2′ and 3–4′. The latter processes reflect the compressor efficiency η_c and turbine efficiency η_t. Besides component efficiencies η_c and η_t, the cycle thermal efficiency η_{th} is very important. The latter efficiency is defined as

$$\eta_{th} = \frac{W_t - W_c}{Q_A}$$

(7.1)

where W_t is turbine work per unit mass of gas, W_c is compressor work, and Q_A is the heat added to the gas in process 2′–3.

Cycle thermal efficiency depends on the cycle pressure ratio P_2/P_1 and the cycle temperature ratio T_3/T_1. Usually, the design value of the peak temperature T_3 is raised as high as possible, consistent with the required life

151

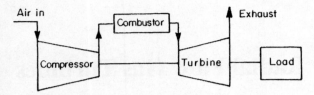

Figure 7.1 Gas-turbine power plant.

of the first-stage blades, and the cycle pressure ratio is chosen as that value corresponding to maximum cycle thermal efficiency or maximum specific output $W_t - W_c$. Thus determination of the optimum cycle pressure ratio is a logical first step in the design of a gas turbine, and it provides the turbomachine designer with the thermodynamic state of the gas as it enters the first stage of the turbine.

Another point that should be made is that the compressor and turbine are interdependent. Usually mounted on the same shaft, the speed of one must be the speed of the other. Furthermore, the pressure ratios and mass flows are also roughly equal. The determination of a point of operation is a process known as matching, and the two machines are matched when speeds, mass flows, and pressure ratios are equal. Both compressor and turbine maps are required to carry out the matching process.

In this chapter we shall consider primarily the design and performance of radial-flow turbines, since this type is widely used in auxilliary power units, gas processing units, turbochargers, turboprop aircraft engines, and waste-heat and geothermal power recovery units. In Chapter 8 the axial-

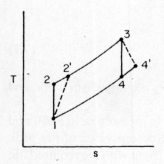

Figure 7.2 Thermodynamic cycle.

flow turbine will be discussed; it is preferred for power plant standby units and large aircraft engines.

Three additional radial-flow turbines should be mentioned. The Francis-type hydraulic turbine, which was first developed in 1847 and is, at present, widely used in various sizes up to 600 MW in hydraulic power plants, is very similar in geometry to the gas turbine described in the present chapter, but it makes use of water as the flowing fluid. Hydraulic turbines are discussed in Chapter 10. The Tesla turbine, which was invented in 1911 by N. Tesla, is a vaneless form of a radial-inflow turbine, which gains its motive force from fluid friction acting on closely spaced disks mounted on a shaft. Rice (1991) presents a critical discussion of recent research of this potentially useful machine. The Ljungstroem turbine, discussed by Shepherd (1956) and Dixon (1975) is a radial-outflow turbine that has been widely used in steam power plants.

7.2   Basic Theory

A typical radial-flow gas turbine is constructed like a centrifugal compressor with radial vanes at the tip of the impeller. However, flow in the turbine is opposite to that of the compressor, i.e., the gas flows from the outer to the inner radius. Figure 7.3 shows the basic elements of the radial-flow turbine. First, the gas enters the volute, which distributes it to the nozzle ring, or stator, located around the casing between the volute and the rim of the rotor. The stator vanes, shown in Figures 7.3 and 7.4, expand the gas to velocity V_2 and direct it to exit at flow angle α_2 just before it impacts on the rotor vanes. The relative velocity W_2 enters the rotor at radial position r_2 and flows radially and then axially through the passages between the rotor vanes. Figure 7.4 indicates than the trailing portion of the rotor vanes are curved, so that the relative velocity W_3 leaves the rotor with a tangential as well as an axial component and makes the angle β_3 with the axial direction. Velocity triangles for rotor inlet and exit are shown in Figure 7.5. The gas exhausts from the rotor at pressure p_3 and undergoes a flow compression in the diffuser, from which it exits at section 4 (Figure 7.3) at atmospheric pressure p_a.

Applying equation (2.9) to the stator results in the energy equation,

$$h_1 + \frac{V_1^2}{2} = h_2 + \frac{V_2^2}{2}$$

(7.2)

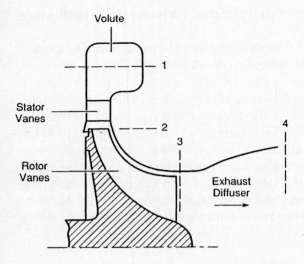

Figure 7.3 Longitudinal section of radial-flow turbine.

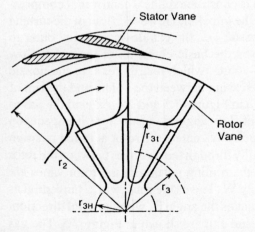

Figure 7.4 Transverse section of radial-flow turbine.

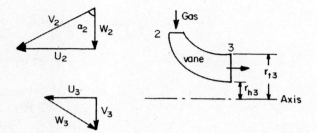

Figure 7.5 Velocity diagrams for radial-flow gas turbines.

Since total properties in the volute are usually known, we can write (7.2) as

$$h_{o1} = h_2 + \frac{V_2^2}{2}$$

(7.3)

Similarly, the energy equation for the exhaust diffuser will be

$$h_3 + \frac{V_3^2}{2} = h_4 + \frac{V_4^2}{2}$$

(7.4)

Applying the adiabatic flow equation (2.9) and the Euler equation (2.16) to the turbine rotor yields

$$h_2 + \frac{V_2^2}{2} = h_3 + \frac{V_3^2}{2} + E$$

(7.5)

and

$$E = V_{u2}U_2 - V_{u3}U_3$$

(7.6)

Although $V_{u3} = 0$ for the usual IFR gas turbine, we will first consider the more general form of the velocity triangles, in which W_2 is not radial and V_3 is not axial. In this case, we can write the following set of equations:

$$V_2^2 = V_{u2}^2 + V_{r2}^2$$

(7.7)

$$V_3^2 = V_{u3}^2 + V_{r3}^2 \tag{7.8}$$

$$W_{u2} = V_{u2} - U_2 \tag{7.9}$$

$$W_{u3} = V_{u3} - U_3 \tag{7.10}$$

$$W_2^2 = W_{u2}^2 + V_{r2}^2 \tag{7.11}$$

$$W_3^2 = W_{u3}^2 + V_{r3}^2 \tag{7.12}$$

Substitution of (7.9) and (7.10) into (7.11) and (7.12), application of (7.8) and (7.9) in the resulting expressions and final substitution for the kinetic energy terms in (7.5), leads to the important result,

$$h_2 + \frac{W_2^2}{2} - \frac{U_2^2}{2} = h_3 + \frac{W_3^2}{2} - \frac{U_3^2}{2} \tag{7.13}$$

If the moving gas is brought to rest with respect to the moving rotor blade at any point in the rotor passage, its enthalpy at that point becomes the relative total enthalpy,

$$h_{oR} = h + \frac{W^2}{2} \tag{7.14}$$

and the corresponding relative total pressure and temperature of the gas are denoted by p_{oR} and T_{oR}. The relative stagnation (or total) states are depicted on the temperature-entropy diagram in Figure 7.6. The temperatures at these states are obtained from (7.14) by dividing by c_p they are

$$T_{o2R} = T_2 + \frac{W_2^2}{2c_p} \tag{7.15}$$

and

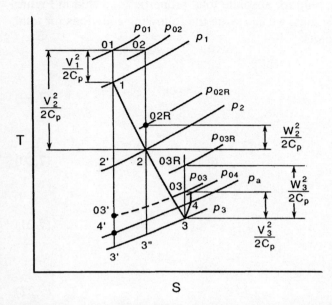

Figure 7.6 Thermodynamic processes in radial-flow turbine.

$$T_{o3R} = T_3 + \frac{W_3^2}{2c_p} \tag{7.16}$$

In Figure 7.6 the points 2 and 02R are on the same constant entropy line; thus,

$$\frac{P_{o2R}}{P_2} = \left(\frac{T_{o2R}}{T_2}\right)^{\gamma/(\gamma-1)} \tag{7.17}$$

Similarly, the isentropic relationship between 3 and o3R warrants the use of

$$\frac{P_{o3R}}{P_3} = \left(\frac{T_{o3R}}{T_3}\right)^{\gamma/(\gamma-1)} \tag{7.18}$$

Similar relations hold for absolute total properties depicted in Figure 7.6 as 01, 02, and 03; thus, we can write the following equations for total properties at these points:

$$h_{o1} = h_1 + \frac{V_1^2}{2} \tag{7.19}$$

$$h_{o2} = h_2 + \frac{V_2^2}{2} \tag{7.20}$$

$$h_{o3} = h_3 + \frac{V_3^2}{2} \tag{7.21}$$

$$T_{o1} = T_1 + \frac{V_1^2}{2c_p} \tag{7.22}$$

$$T_{o2} = T_2 + \frac{V_2^2}{2c_p} \tag{7.23}$$

$$T_{o3} = T_3 + \frac{V_3^2}{2c_p} \tag{7.24}$$

$$p_{o1} = p_1 \left(\frac{T_{o1}}{T_1} \right)^{\gamma/(\gamma - 1)} \tag{7.25}$$

$$p_{o2} = p_2 \left(\frac{T_{o2}}{T_2} \right)^{\gamma/(\gamma - 1)} \tag{7.26}$$

$$p_{o3} = p_3 \left(\frac{T_{o3}}{T_3} \right)^{\gamma/(\gamma - 1)} \tag{7.27}$$

For the 90° IFR gas turbine (inflow radial with $W_2 = V_{r2}$), the basic equations can be simplified considerably. For example, since $V_{u2} = U_2$ and $V_{u3} = 0$,

$$E = U_2^2 \tag{7.28}$$

$$W_2^2 = U_2^2 \cot^2 \alpha_2 \tag{7.29}$$

$$V_3^2 = U_3^2 \cot^2 \beta_3 \tag{7.30}$$

and

$$W_3^2 = V_3^2 + U_3^2 \tag{7.31}$$

Substituting the above equations into (7.13) yields the equation for rotor temperature ratio,

$$\frac{T_3}{T_2} = 1 - \frac{(\gamma - 1)U_2^2}{2a_2^2}\left(1 - \cot^2 \alpha_2 + \frac{r_3^2}{r_2^2}\cot^2 \beta_3\right) \tag{7.32}$$

Referring to Figure 7.6, a direct isentropic process exists between the stagnation state 01, defined by p_{o1} and T_{o1}, and the state of the flowing rotor exhaust, defined by p_3 and $T_{3'}$. The ideal expansion 01 to 3' could take place in an ideal turbine or in an ideal nozzle. The turbine process can be idealized further by imagining that the exhaust gas leaves the turbine with zero kinetic energy, i.e., $V_{3'} \to 0$; thus, $h_{o3'} \to h_{3'}$, and equation (7.5), when applied to the ideal turbine, becomes

$$h_2 + \frac{V_2^2}{2} = h_{3'} + \frac{V_{3'}^2}{2} + E_i \tag{7.33}$$

Noting that $V_{3'} \to 0$ and applying (7.3), we have

$$E_i = h_{o1} - h_{3'} \tag{7.34}$$

The ideal process from state 01 to state 3', as previously noted in Figure 7.6, can also take place in an ideal nozzle. In this case, the exit velocity becomes the maximum possible nozzle exit velocity c_o, which is called the spouting velocity. Adapting equation (7.3) to the ideal nozzle, we can write

$$\frac{c_o^2}{2} = h_{o1} - h_{3'} \tag{7.35}$$

Equating the left hand sides of (7.34) and (7.35), we obtain

$$E_i = \frac{c_o^2}{2} \tag{7.36}$$

Substituting for E_i with equation (7.28), we find that

$$U_{2i} = 0.707c_o \tag{7.37}$$

which gives the rotor tip speed for the ideal 90° IFR gas turbine. The turbine efficiency of the ideal turbine is 100 percent, whether the total-to-total efficiency or the total-to-static efficiency is used.

The total-to-static efficiency is defined as

$$\eta_{ts} = \frac{h_{o1} - h_{o3}}{h_{o1} - h_{3'}} \tag{7.38}$$

It is clear from equations (7.3) and (7.5) that the numerator of (7.38) represents the energy transfer E of the actual turbine, whereas equation (7.34) shows that the denominator of equation (7.38) represents the energy transfer E_i of an ideal turbine. An alternative form of equation (7.38) is obtained through the employment of equation (7.36); thus we can write

$$\eta_{ts} = \frac{2E}{c_o^2} \tag{7.39}$$

Practically, equation (7.37) permits the designer to estimate the upper limit of rotor tip speed for a given set of inlet and exit conditions. This is apparent from the spouting velocity equation,

$$c_o = \left\{ \frac{2\gamma RT_{o1}}{\gamma - 1} \left[1 - \left(\frac{p_3}{p_{o1}} \right)^{(\gamma - 1)/\gamma} \right] \right\}^{1/2}$$

(7.40)

which can be derived from (7.35).

The total-to-total efficiency of a turbine is defined as

$$\eta_{tt} = \frac{h_{o1} - h_{o3}}{h_{o1} - h_{o3'}}$$

(7.41)

where

$$h_{o3'} = h_{3'} + \frac{V_{3'}^2}{2}$$

(7.42)

Figure 7.6 shows that $p_{o3'} = p_{o3}$ and $p_{3'} = p_3$. Additionally, it is seen that states 03 and 3, 03' and 3', and 01 and 3' are connected by isentropic processes; thus, we can write

$$\frac{T_{o3}}{T_{3'}} = \left(\frac{p_{o3'}}{p_{3'}} \right)^{(\gamma - 1)/\gamma}$$

(7.43)

$$\frac{T_{o3}}{T_3} = \left(\frac{p_{o3}}{p_3} \right)^{(\gamma - 1)/\gamma}$$

(7.44)

$$\frac{T_{o1}}{T_{3'}} = \left(\frac{p_{o1}}{p_3} \right)^{(\gamma - 1)/\gamma}$$

(7.45)

The equality of pressure ratios means that

$$\frac{T_{o3'}}{T_{3'}} = \frac{T_{o3}}{T_3}$$

(7.46)

Dividing numerator and denominator of (7.41) by c_p gives

$$\eta_{tt} = \frac{T_{o1} - T_{o3}}{T_{o1} - T_{o3'}}$$

$$(7.47)$$

$T_{o3'}$ for the above equation is found from (7.45) and (7.46). Dividing (7.42) by c_p yields

$$T_{o3'} = T_{3'} + \frac{V_{3'}^2}{2c_p}$$

$$(7.48)$$

The above equation can be used to determine $V_{3'}$.

Combining (7.5), (7.38), (7.42), (7.47), and (7.48) produces

$$\eta_{tt} = \frac{1}{(1/\eta_{ts}) - (V_{3'}^2)/(2E)}$$

$$(7.49)$$

which allows the calculation of one efficiency from the other.

Combining (7.3), (7.5), (7.21), (7.24), and (7.38) results in

$$\eta_{ts} = \frac{E}{E + c_p(T_3 - T_{3'}) + V_3^2/2}$$

$$(7.50)$$

Assuming that U_2, U_3, β_2, and γ_3 are known, and that the volute pressure and temperature, p_{o1} and T_{o1}, and the turbine exhaust pressure p_3 are also known, the efficiency η_{ts} can be calculated from (7.50).

The turbine exhaust total temperature T_{o3} can be found from (7.5), which can be written as

$$T_{o3} = T_{o1} - \frac{E}{c_p}$$

$$(7.51)$$

The static turbine exhaust temperature T_3 is determined from (7.24), and the ideal exhaust temperature $T_{3'}$ is calculated from (7.45). Substituting these values into (7.50) yields η_{ts}.

If only U_2, U_3, and α_2 are available, then the loss coefficients λ_N and λ_R must also be known in order to determine the efficiency.

The nozzle loss coefficient is defined as

$$\lambda_N = \frac{2c_p(T_2 - T_{2'})}{V_2^2}$$

(7.52)

and is used to calculate $T_{2'}$ (see Figure 7.6). Since $p_2 = p_{2'}$, we can write

$$p_2 = p_{o1}\left(\frac{T_{2'}}{T_{o1}}\right)^{\gamma/(\gamma - 1)}$$

(7.53)

Since $p_3 = p_{3'}$, the isentropic relation,

$$T_{3''} = T_2\left(\frac{p_3}{p_2}\right)^{(\gamma - 1)/\gamma}$$

(7.54)

is used to find $T_{3''}$. The rotor loss coefficient λ_R is defined as

$$\lambda_R = \frac{2c_p(T_3 - T_{3''})}{W_3^2}$$

(7.55)

and serves as one relationship between β_3 and T_3, viz.,

$$T_3 = T_{3''} + \lambda_R \frac{U_3^2 \csc^2 \beta_3}{2c_p}$$

(7.56)

where $U_3 \csc \beta_3$ is substituted for W_3. A second relationship is provided by (7.24) and (7.30), viz.,

$$T_3 = T_{o3} - \frac{U_3^2 \cot^2 \beta_3}{2c_p}$$

(7.57)

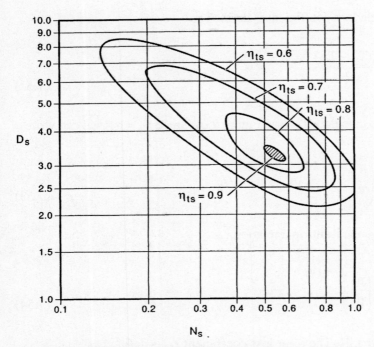

Figure 7.7 Balje diagram for 90° radial-inflow gas turbines.

Sources: Scheel (1972) and Whitfield and Baines (1990)

The above equations can be solved simultaneously for T_3 and β_3. Finally, V_3^2 is calculated from (7.30), and the efficiency η_{ts} is determined from (7.50).

An alternative to calculation of η_{ts} from the loss coefficients, as is described above, is the use of the Balje diagram, which consists of iso-efficiency curves plotted as functions of specific speed and specific diameter. The Balje diagram for the 90° IFR gas turbines is presented in Figure 7.7. The efficiencies of Figure 7.7 are conservative, except for the shaded area in the center, which, according to Whitfield and Baines (1990), represents designs with efficiencies of 0.9.

The Balje diagram can be used by the designer to select combinations of N_s and D_s which correspond to a given efficiency. Subject to stress limitations, the rotor tip diameter can be determined from the chosen specific diameter. It should be noted that the spouting velocity appears in the definitions of both specific diameter and specific speed; thus,

$$D_s = \frac{D_2[(c_o^2)/2]^{1/4}}{Q_3^{1/2}}$$

$$(7.58)$$

$$N_s = \frac{NQ_3^{1/2}}{[(c_o^2)/2]^{3/4}}$$

$$(7.59)$$

where c_o can be determined from (7.40), and Q_3 is the volume flow rate of gas measured at exhaust conditions. The latter flow rate is determined from the exhaust density and the mass flow rate using

$$Q_3 = \frac{\dot{m}}{\rho_3}$$

$$(7.60)$$

When (7.58) and (7.59) are multiplied together, their product is

$$N_s D_s = \frac{(2^{3/2})U_2}{c_o}$$

$$(7.61)$$

Thus, the Balje diagram is a useful design tool for the initial selection of η_{ts}, D_s, N_s, and U_2/c_o. The details of the design process is considered in the next section.

7.3 Design

The Balje diagram provides a starting point in the design process. Using a Balje diagram, e.g., Figure 7.7, allows choices to be made of N_s, D_s, and η_{ts}. Equation (7.61) can be used to determine the corresponding value of U_2/c_o, which can then be checked against recommended design values for this parameter (see Table 7.1). If p_{o1}, T_{o1}, and p_3 are known, (7.40) can be used to determine c_o, and, finally, U_2 is found from the ratio U_2/c_o and checked against the structural limit for U_2 set by a stress analyst, e.g., 1600–1700 fps is a possible range of acceptable values for tip speed.

Figure 7.8 depicts the counterrotating relative eddies in the blade passages of the 90° IFR gas turbine. Whitfield and Baines (1990) show that negative incidence, as shown in the velocity diagram of Figure 7.8,

Table 7.1 Design Parameters for 90° IFR Gas Turbines

Parameter	Recommended range	Source
α_2	68–75°	Dixon, Rohlik
β_3	50–70°	Whitfield & Baines
D_{3H}/D_{3S}	<0.4	Dixon, Rohlik
D_{3S}/D_2	<0.7	Dixon, Rohlik
D_3/D_2	0.53–0.66	Whitfield & Baines
b_2/D_2	0.05–0.15	Whitfield & Baines, Dixon, Rohlik
U_2/c_0	0.55–0.80	Figure 7.7
W_3/W_2	2–2.5	Ribaud & Mischell
V_3/U_2	0.15–0.5	Whitfield & Baines
λ_R	0.4–0.8	Dixon
λ_N	0.06–0.24	—

strengthens the relative eddies; the result is increased vane pressure difference, energy transfer, and efficiency. The optimum value of incidence β_2 for maximizing this effect is given as

$$\cos \beta_2 = 1 - \frac{\pi 0.63}{n_B} \tag{7.62}$$

where n_B is the number of vanes. The minimum number of vanes to prevent flow reversal is given by Glassman (1976) as

$$n_B = 0.1047(110 - \alpha_2) \tan \alpha_2 \tag{7.63}$$

where α_2 is in degrees.

If zero incidence is selected, then (7.28) is used to determine E. With negative incidence, E can be determined from (7.6) with $V_{u3} = 0$ and

$$V_{u2} = U_2 - V_{m2} \tan \beta_2 \tag{7.64}$$

where

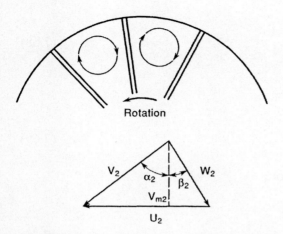

Figure 7.8 Relative eddies in IFR turbines.

$$V_{m2} = \frac{U_2}{\tan \beta_2 + \tan \alpha_2}$$

(7.65)

Once the energy transfer has been determined, the mass flow of gas required to achieve the specified power can be calculated from (2.15), which can be expressed as

$$\dot{m} = \frac{P}{E}$$

(7.66)

D_{3S} and D_{3H} are determined from diameter ratios which are selected from the ranges given in Table 7.1 and later modified to achieve compliance with recommended ranges of values of β_3 and V_3/U_2 recommended in Table 7.1. The rotor exit velocity V_3 and the axial width b_2 are determined from (2.5) using

$$V_3 = \frac{\dot{m}}{\rho_3 A_3}$$

(7.67)

and

$$b_2 = \frac{\dot{m}}{W_2 \rho_2 \pi D_2}$$
(7.68)

where

$$A_3 = \frac{\pi(D_{3S}^2 - D_{3H}^2)}{4}$$
(7.69)

The angle β_3 is determined from

$$\beta_3 = \tan^{-1} \frac{U_3}{V_3}$$
(7.70)

The exit blade velocity U_3 in (7.70) is based on the rms mean value of diameter, i.e.,

$$D_3 = \left(\frac{D_{3S}^2 + D_{3H}^2}{2} \right)^{1/2}$$
(7.71)

Finally, the ratios b_2/D_2 and W_3/W_2 are checked for agreement with recommended ranges in Table 7.1.

7.4   Examples

Example Problem 7.1

Design a 90° IFR gas turbine which produces 700 hp while running at 41,000 rpm with a turbine inlet temperature T_{o1} of 1700°R and an exhaust pressure p_3 of 14.3 psia. Assume zero exhaust swirl and zero incidence. Take $\gamma = 1.35$, $R = 1714$ ft-lb/Sl-°R and $c_p = 6611$ ft-lb/Sl-°R.
Solution: Select a point on Figure 7.7 where $N_s = 0.67$, $D_s = 2.7$, and $\eta_{ts} = 0.77$. Using (7.61), the speed ratio is

$$\frac{U_2}{c_o} = 0.64$$

Choosing $U_2 = 1600$ fps, then $c_o = 2500$ fps. Solving (7.40) for the required volute pressure, we have $p_{o1} = 50.2$ psia. The energy transfer is obtained from (7.28) and is

$$E = (1600)^2 = 2,560,000 \ \text{ft}^2/\text{s}^2$$

The mass flow rate is determined from (7.66) and is

$$\dot{m} = \frac{700(550)}{2,560,000} = 0.15 \ \text{Sl/s} = 4.84 \ \text{lb/s}$$

The tip diameter is obtained from (2.1); thus,

$$D_2 = \frac{2U_2}{N} = \frac{2(1600)}{4293.51} = 0.745 \ \text{ft}$$

The tip diameter is rounded to 0.75 ft or 9 inches. Choose $D_{3S} = 0.75D_2 = 6.75$ inches. This is a little higher than the rule of Table 7.1, but appears to be necessary to reduce the exhaust velocity V_3. Choose $\alpha_2 = 72.5°$ and calculate W_2 using (7.29); thus,

$$W_2 = U_2 \cot \alpha_2 = 505 \ \text{fps}$$

Also

$$V_2 = U_2 \csc \alpha_2 = 1678 \ \text{fps}$$

From (7.23) we have

$$T_2 = 1487°R$$

from which $a_2 = 1855$ fps and $M_2 = 0.904$. Choosing $\lambda_N = 0.1$ and applying (7.52), we find

$$T_{2'} = 1487 - \frac{0.1(1678)^2}{2(6611)} = 1466°R$$

From (7.53) we find that $p_2 = 28.4$ psia, and the corresponding density is 0.052 lb/cu ft. Equation (7.68) is used to calculate b_2, which is 0.94 inch. Thus,

$$\frac{b_2}{D_2} = \frac{0.94}{9} = 0.104$$

which is an acceptable value. Choose $D_{3H} = 2.5$ inches, which gives an acceptable hub-tip ratio of 0.37. The rms mean diameter, calculated from (7.71), is 5.09 inches. This yields a diameter ratio of

$$\frac{D_3}{D_2} = \frac{5.09}{9} = 0.565$$

which is an acceptable value. The corresponding U_3 is 911 fps. Utilizing (7.51), (7.24), and (7.67) gives $V_3 = 743$ fps. From (7.70) the relative flow angle at the exit is

$$\beta_3 = \tan^{-1} \frac{911}{743} = 51°$$

which is in the acceptable range. From (7.31) the relative velocity at the exit is $W_3 = 1176$ fps. The relative velocity ratio is

$$\frac{W_3}{W_2} = \frac{1176}{505} = 2.33$$

which is also acceptable, according to Table 7.1. Finally, the number of vanes is calculated from (7.63):

$$n_B = 0.1047(110 - 72.5) \tan 72.5 = 12.45$$

Choose $n_B = 12$ vanes.

References

Dixon, S. L. 1975. *Fluid Mechanics. Thermodynamics of Turbomachinery*. Pergamon, Oxford.

Glassman, A. J. 1976. *Computer Program for Design and Analysis of Radial Inflow Turbines*. NASA TN 8164.

Ribaud, Y., and C. Mischell. 1986. *Study and Experiments of a Small Radial Turbine for Auxiliary Power Units*. TP No. 1986-55. ONERA, Chatillon.

Rice, W. 1991. Tesla Turbomachinery. *Proceeding of the IV International Tesla Symposium*. Serbian Academy of Sciences and Arts, Belgrade.

Rodgers, C., and R. Geiser. 1987. Performance of High-Efficiency Radial/Axial Turbine. *Journal of Turbomachinery*. ASME, New York.

Rohlik, H. E. 1968. *Analytical Determination of Radial Inflow Turbine Design Geometry for Maximum Efficiency*. NASA TN D-4384.

Scheel, L. F. 1972. *Gas Machinery*. Gulf Publishing Co., Houston.

Shepherd, D. G. 1956. *Principles of Turbomachinery*. Macmillan, New York.

Whitfield, A., and N. C. Baines. 1990. *Design of Radial Turbomachines*. Longman, Essex.

Problems

7.1 A radial-flow gas turbine with flat radial vanes is to run at 24,200 rpm. Gas is to enter the rotor at a radius of 6 in. and exit at a mean radius of 3 in. Exhaust gases are to leave at 14.7 psia and 700°F and are to have a relative Mach number of 0.75. Calculate the mass flow rate to produce 100 hp, and estimate the blade height at the exit.

7.2 Derive equations (7.13) and (7.32).

7.3 Derive equation (7.40) for spouting velocity.

7.4 Derive equation (7.50).

7.5 A 90° IFR gas turbine with flat radial blades runs at 24,200 rpm. The radius of the rotor at the inlet is 4 inches, and the mean radius at the exit is 2.25 inches. Compressor bleed air at a (total) temperature of 700°R is supplied. The exhaust (static) pressure is 14.7 psia. The nozzle angle is 70 degrees. Assume zero incidence and deviation. Find:

        a) the mass flow rate of air required to produce 100 horse power

        b) the nozzle exit static temperature

        c) the exhaust total temperature

7.6 Assume a velocity coefficient of 0.95, φ = velocity coefficient = $V_2/V_{2'}$ and a rotor loss coefficient of 1.0 for the turbine in Problem 7.5. $b_3/b_2 = 1.925$. Find:

    a) the mean exit gas angle with respect to axial
    b) the spouting velocity
    c) total-to-static efficiency
    d) total-to-total efficiency
    e) the total pressure in the scroll
    f) find the blade height at the inlet and exit

7.7 For the turbine of Problems 7.5 and 7.6 the blades are extended to turn the flow sufficiently to eliminate the exhaust swirl. The blade heights are as in 7.6(f). The magnitude of the velocity at the nozzle exit remains fixed. Find:

    a) the mean exit blade angle required
    b) the new total-to-static efficiency
    c) the new scroll total pressure
    d) the new turbine horse power
    e) the rotor loss coefficient

7.8 A standard 90° IFR gas turbine runs at 41,000 rpm. The rotor tip diameter is 8.5714 inches and the tip vane height is 0.875 inch. Gas leaves the nozzle at an angle of 68° and a Mach number of 1.0. The total-to-static efficiency is 0.80, and the nozzle velocity coefficient is 0.95, φ = velocity coefficient = $V_2/V_{2'}$. The exhaust pressure is one atmosphere. Find:

    a) the relative velocity entering the rotor
    b) the static temperature at the rotor inlet
    c) the total pressure at the nozzle inlet
    d) the static pressure at the rotor inlet
    e) the power produced in horsepower

7.9 A standard 90° IFR gas turbine runs at 60,000 rpm during a performance test. The throttle total temperature $T_{o1} = 1800°R$, and the turbine pressure ratio $P_{o1}/P_3 = 2.2$. The combined mass rate of flow of the air and fuel is 0.71 lb/s. The tip diameter D_2 is 5 inches. Assume that the mechanical efficiency is 95 percent, that the ratio of specific heats of the gas is 1.35 and that the exhaust swirl is completely removed by the exducer. Determine:

    a) the ratio of tip speed to spouting velocity
    b) the brake horsepower produced by the turbine
    c) the total-to-static efficiency

7.10 It is required to determine the rotor loss coefficient and the total-to-static efficiency from available test data from a test of a 90° IFR gas turbine. The test data are the following: brake power = 139.1 hp; fuel flow rate = 209 lb/hr; air flow rate = 4.85 lb/s; ambient air temperature = 73°F; ambient air pressure = 14.27 psia; compressor discharge pressure = 45 psig, compressor discharge temperature = 411°F; exhaust gas total temperature = 880°F; rotor speed = 41,000 rpm; rotor tip diameter = 9.1 in.; blade tip width = .85 in.; rotor exit tip diameter = 6.75 in.; exit hub diameter = 2.5 in. The following assumptions will be made as well: nozzle velocity coefficient φ = .95; ratio of specific heats for the turbine gas = 1.35; three percent of the total pressure is lost in the burner and 5.4 percent of the turbine energy transfer is lost in disk and bearing friction and through power takeoffs to drive auxiliaries. Find:

a) turbine power from compressor and brake power
b) turbine power from E and mass flow
c) nozzle angle to produce required mass flow
d) pressure at nozzle exit
e) temperature at nozzle exit
f) mean flow angle at rotor exit
g) rotor loss coefficient
h) total-to-static efficiency

7.11 Use the methods developed in the text to design a single-stage 90° IFR gas turbine rotor which satisfies the following requirements:

Speed = 41,000 rpm
Power = 700 hp
Turbine Inlet (total) Temperature = 1700°R
Turbine Inlet (total) Pressure = 57.5 psia
Turbine Exhaust (static) Pressure = 14.3 psia

Symbols for Chapter 7

a	local acoustic speed in gas
a_2	acoustic speed at rotor inlet
A_3	annular flow area at rotor exit
b_2	width of vane at $r = r_2$
c_o	spouting velocity
c_p	specific heat at constant pressure

D_{3H} hub diameter at rotor outlet
D_{3S} shroud diameter at rotor outlet
D_2 rotor tip diameter
D_s specific diameter
E energy transfer from fluid to rotor
E_i energy transfer for isentropic turbine
h_1 specific enthalpy of gas in volute
h_2 specific enthalpy of gas entering rotor
h_3 specific enthalpy of gas leaving rotor
$h_{3'}$ specific enthalpy of gas leaving ideal rotor
h_4 specific enthalpy of gas leaving diffuser
h_{o1} total enthalpy of gas entering stator
h_{o2} total enthalpy of gas leaving stator
h_{o3} total enthalpy of gas leaving rotor
$h_{o3'}$ total enthalpy of gas leaving ideal rotor
h_{oR} relative total enthalpy
M_2 absolute Mach number at rotor inlet $= V_2/a_2$
$\dot{m}$ mass flow rate of gas
N rotor speed
N_s specific speed
n_B number of vanes (or blades) in the rotor
P turbine power
p_1 static pressure of gas in volute
p_2 static pressure of gas leaving stator
$p_{2'}$ static pressure of gas leaving ideal stator
p_3 static pressure of gas leaving rotor
$p_{3'}$ static pressure of gas leaving ideal rotor
p_{o1} total pressure of gas in volute
p_{o2} total pressure of gas at stator outlet
p_{o3} total pressure of gas at rotor outlet
$p_{o3'}$ total pressure of gas leaving ideal rotor
p_{o2R} relative total pressure leaving stator
p_{o3R} relative total pressure leaving rotor
Q_3 volume flow rate based on gas density at rotor outlet
Q_A heat added in gas turbine cycle
r radial position measured from axis of rotation
r_2 radial position at tip of rotor $= D_2/2$
r_3 rms mean radial position at exit from rotor $= D_3/2$
R gas constant $= R_u/M$

R_u	universal gas constant
T	gas temperature
T_o	total temperature of gas
T_1	static temperature of gas in volute
T_2	static temperature of gas leaving stator
$T_{2'}$	static temperature of gas leaving ideal stator
T_3	static temperature of gas leaving rotor
$T_{3'}$	static temperature of gas leaving ideal turbine
$T_{3''}$	static temperature of gas leaving ideal rotor with inlet temperature T_2
T_{o1}	total temperature of gas in volute
T_{o2}	total temperature of gas leaving stator
T_{o3}	total temperature of gas leaving rotor
T_{o2R}	relative total temperature of gas leaving stator
T_{o3R}	relative total temperature of gas leaving rotor
U_2	tip speed of rotor
U_{2i}	tip speed of ideal rotor
U_{3H}	U at rotor exit at hub diameter
U_{3S}	U at rotor exit at shroud diameter
U_3	U at rotor exit at rms mean diameter
V_1	absolute velocity of gas in volute
V_2	absolute velocity of gas at stator exit
V_{m2}	meridional component of $V_2 = V_{r2}$
V_{r2}	radial component of V at $r = r_2$
V_{u2}	tangential component of V_2
V_3	absolute velocity of gas leaving the rotor
V_{a3}	axial component of V_3
V_{m3}	meridional component of $V_3 = V_{a3}$
V_{u3}	tangential component of V_3
$V_{3'}$	absolute velocity of gas at exit of ideal turbine
V_4	absolute velocity of gas at exit of diffuser
W	velocity relative to moving vane
W_2	relative velocity of gas at tip of rotor
W_{m2}	meridional component of W_2
W_{u2}	tangential component of W_2
W_3	relative velocity of gas leaving the rotor
W_{u3}	tangential component of W_3
W_c	compressor work
W_t	turbine work
α_2	$\tan^{-1}(V_{u2}/V_{m2})$; absolute gas angle at r

α_3 absolute gas angle at $r = r_3$
β_2 angle between W_2 and V_{m2}
β_3 angle between W_3 and V_{m3}
γ ratio of specific heats
η_{ts} total-to-static turbine efficiency
η_{tt} total-to-total turbine efficiency
η_{th} thermal efficiency of gas turbine cycle
λ_N nozzle loss coefficient
λ_R rotor loss coefficient
φ nozzle velocity coefficient $V_2/V_{2'}$
ρ gas density
ρ_2 gas density leaving stator
ρ_3 gas density leaving rotor

8 Axial-Flow Gas Turbines

8.1 Introduction

Besides the radial-flow gas turbine, which is discussed in Chapter 7, two additional types of gas turbines are commonly used, viz., the axial-flow impulse turbine and the axial-flow reaction turbine. The axial-flow impulse turbine is discussed in Chapter 2 and in Chapter 9. As shown in Figure 2.5, the impulse turbine has a symmetrical blade profile; moreover, the velocity diagram shown in Figure 9.2 is drawn with $W_2 = W_3$, i.e., there is no change in the magnitude of the relative velocity of the gas in the impulse rotor. The reaction equation (2.23), when applied to the impulse rotor with inlet at station 2 and exit at station 3, shows that $R = 0$, i.e., the impulse turbine is a zero reaction turbomachine. Equation (9.18) shows further that $h_2 = h_3$, which, for a perfect gas, means that $T_2 = T_3$ and $p_2 = p_3$. In the stator or nozzle section, which is located upstream of the rotor, the gas expands and its velocity increases. Application of equations (2.9) and (2.10) to the impulse rotor yields

$$E = h_{o2} - h_{o3} \tag{8.1}$$

and

$$E = \frac{V_2^2}{2} - \frac{V_3^2}{2} \tag{8.2}$$

177

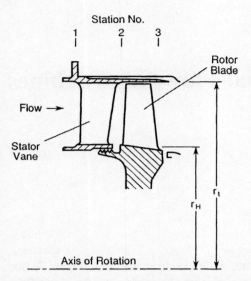

Figure 8.1 Sectional view of typical axial-flow turbine stage.

The latter equation is the same as (2.19) applied to an impulse rotor having stations numbered 2 and 3 at the inlet and exit, respectively.

From the above description of an impulse stage, it is clear that the rotor reduces the gas kinetic energy, which is created in the stator. The Rateau, or pressure-compounded turbine, comprises multiple impulse stages of the type described above. The Curtis, or velocity-compounded turbine, utilizes multiple passes through rotor blades, which are in series downstream of a single stator; each pass of the gas through a blade row causes a further extraction of kinetic energy from the gas. The Curtis stage is commonly found in steam turbines and is discussed in Chapter 9.

The reaction turbine, i.e., a turbine having a stage reaction > 0, is commonly found in multistage gas turbines used in stationary power plants, as well as in gas turbine engines used to drive ships, trains, and aircraft. A single stage of a reaction turbine comprises a row of stator vanes followed by a row of rotor blades arranged as shown in Figure 8.1.

Reaction blades of any length are commonly twisted as shown in Figure 8.2. Changes in blade profile, such as are shown in Figure 8.2, are made to create a free vortex distribution in the tangential component of the absolute velocity; thus, the product $V_u r$ is constant at station 2 and at station 3 from the hub to the blade tip. This means that $U\Delta V_u$ and hence E are constant as

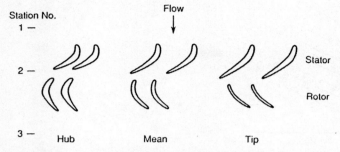

Figure 8.2 Blade profiles in typical axial-flow turbine stage.

well, so that each parcel of gas passing through the rotor loses the same energy per unit mass. The condition is a convenient one for preliminary design, since a single radial position, viz., the rms mean radius, can be used to determine the velocity diagram, the energy transfer, and the efficiency, and the results can be applied to the entire gas flow passing through a given rotor; this is the so-called mean line analysis.

The symmetrical velocity diagrams shown in Figures 8.3 and 8.4 can be used to represent the mean line diagrams of a typical reaction stage of gas turbine. Because of symmetry, $V_3 = W_2$ and $V_2 = W_3$, and the degree of reaction equals one half. The energy equation (7.6), which applies to axial as well as radial flow gas turbines, reduces to

$$E = U^2 \tag{8.3}$$

for the diagram of Figure 8.4. A comparable impulse turbine velocity diagram could be formed from Figure 8.3, if β_2 is drawn equal in magnitude to β_3, and α_3 is reduced to zero; for this diagram $W_{u2} = U$ and $W_{u3} = -U$ and

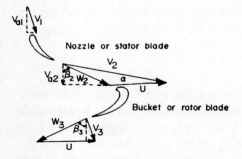

Figure 8.3 Symmetrical diagram.

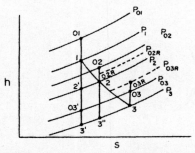

Figure 8.4 Construction of velocity diagrams.

$$E = 2U^2 \tag{8.4}$$

This shows that an impulse stage produces twice the power of a reaction stage with $R = \frac{1}{2}$, if the blade speeds are the same. Although more reaction stages than impulse stages are required to produce the same power, and the reaction turbine is therefore heavier and more expensive, it can be more efficient over a wider range of relatively high blade speeds than the impulse turbine. It is shown in Chapter 9 that the blade speed at which reaction turbines achieve their maximum efficiencies is nearly twice that of impulse turbines, i.e., the optimum value of speed ratio U/c_0 is around $\frac{1}{2}$ for impulse turbines and close to unity for 50 percent reaction turbines. In large multistage turbines impulse blading is often found on the same shaft with reaction blading, i.e., impulse blades are mounted near the inlet where blade length and speed are minimal, while reaction blades are used for most of the later stages where blade size and speed grow with gas expansion and flow area enlargement. The combination of the two types of blades in the same turbine maximizes power and efficiency.

This chapter emphasizes the performance and design of the axial-flow reaction turbine stage. Specifically, the objective of the chapter is to provide the reader with a basis for the prediction of stage efficiency and for the determination rotor and stator dimensions for a specified stage output.

8.2 Basic Theory

A gas turbine stage is shown schematically in Figure 8.5. A row of stationary blades receives the gas in a nearly axial path and deflects it to a small nozzle exit angle α of, say, 15°. The exit velocity V_2 is much larger than the inlet velocity V_1, and the axial component V_{a1} at the inlet is less than that at the exit, i.e., V_{a2}. A near-isentropic expansion of the gas occurs in the nozzle

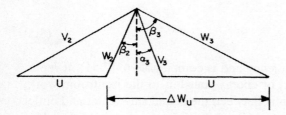

Figure 8.5 Velocity diagrams for a gas turbine.

formed between the blades. The rotor blade, or bucket, turns the gas through a large angle, say 75°. This large change of direction of the relative velocity has a tangential component ΔW_u which gives rise to the large tangential force on the turbine blade.

The key thermodynamic states within a single stage are indicated in Figure 8.6. The actual end states are 1 and 2 in the stator process and 2 and 3 in the rotor process. The corresponding total property states, i.e., 01, 02, and 03, are found by constructing isentropic processes between the actual states and the corresponding total pressure lines (isobars). The states 2′, 03′, and 3′ are those corresponding to an ideal (isentropic) expansion from the inlet to the exit pressure for the stage. The stage efficiency η_s can now be defined as

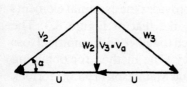

Figure 8.6 Enthalpy-entropy diagram.

$$\eta_{tt} = \frac{h_{o1} - h_{o3}}{h_{o1} - h_{o3'}} \qquad (8.5)$$

which is the total-to-total efficiency and is comparable to (7.41). It represents the ratio of actual work, or energy transfer, to the isentropic work; attainable when the expansion is between the actual stage inlet and outlet total pressures. The total-to-static efficiency is also used and is defined by

$$\eta_{ts} = \frac{h_{o1} - h_{o3}}{h_{o1} - h_{3'}} \qquad (8.6)$$

and is comparable to (7.38).

In order to solve for stage efficiency, it is necessary to establish the velocity diagram. This is carried out through the use of known upstream conditions, p_1, T_1, and V_1; the specified N, P, and $\dot{m}$; and the maximum blade tip speed Nr_t. The tip radius is chosen, so that the tip speed constraint is satisfied. Iteration on the hub-tip ratio follows until the mass flow requirement is satisfied. The hub radius is calculated from the selected hub-tip ratio, and the mean radius is calculated from

$$r_m = \left(\frac{r_t^2 + r_H^2}{2} \right)^{1/2} \qquad (8.7)$$

The mean blade speed U, which is determined from the mean radius, is used to find ΔW_u, which is equal to ΔV_u, using the Euler equation,

$$\Delta W_u = \frac{E}{U} \qquad (8.8)$$

Fixing α_3, which is usually set close to zero, fixes β_3, and β_2 can be determined to satisfy (8.8). The above work provides the essential elements for the construction of a velocity diagram like that of Figure 8.3. The velocity diagram at the mean radius can be used together with empirical loss coefficients to determine the stage efficiency. The method for obtaining valid loss coefficients is described in Ainley and Mathieson (1955), Horlock (1973), Kacker and Okapuu (1982) and Cohen et al. (1987). It is known as the Ainley-Mathieson method.

The rotor loss coefficient Y_R and the stator loss coefficient Y_s are found by adding profile, secondary-flow, and tip-clearance coefficients, i.e.,

$$Y_R = Y_{mR} + Y_{sR} + Y_{tR} \tag{8.9}$$

and

$$Y_s = Y_{ms} + Y_{ss} \tag{8.10}$$

where Y_m is the profile loss coefficient, Y_s is the secondary-flow loss coefficient, and Y_t is the tip-clearance loss coefficient.

The profile loss coefficients are obtained from cascade wind tunnel tests in which total pressure losses are measured, and the results are summarized in Figures 8.7 and 8.8. The stator profile loss coefficient Y_{ms} is read directly

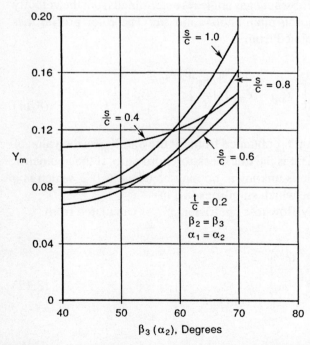

Figure 8.7 Profile loss coefficient for impulse blading.

Source: Cohen et al. (1987)

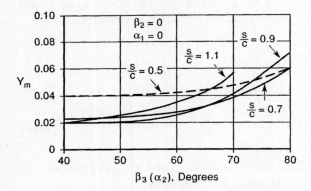

Figure 8.8 Profile loss coefficient for blading with zero inlet gas angle.

Source: Cohen et al. (1987)

from Figure 8.8 with the absolute gas angle α_2 determined from the velocity diagram and with a chosen pitch-chord ratio s/c. The rotor profile loss coefficient Y_{mR} is calculated from

$$Y_{mR} = \left[Y_{mo} + \left(\frac{\beta_2}{\beta_3} \right)^2 (Y_{mI} - Y_{mo}) \right] \left(\frac{5t}{c} \right)^{\beta_2/\beta_3} \tag{8.11}$$

where Y_{mo} is the value of Y_m obtained from Figure 8.8, Y_{mI} is the value of Y_m from Figure 8.7 and t/c is the thickness-to-chord ratio. If the geometric form of the blade profile is unknown, t/c may be taken as 0.2, which is a reasonable value and one which simplifies (8.11).

The stator secondary-flow loss coefficient Y_{ss} is calculated from

$$Y_{ss} = \lambda_s \left(\frac{c_L}{s/c} \right)^2 \left(\frac{\cos^2 \alpha_2}{\cos^3 \alpha_m} \right) \tag{8.12}$$

in which

$$c_L = \frac{2s}{c} (\tan \alpha_2 - \tan \alpha_1) \cos \alpha_m \tag{8.13}$$

and

$$\lambda_s = 0.0996x^{2.1854} + 0.0056 \qquad (8.14)$$

for which

$$x = \frac{R_A^2}{1+k} \qquad (8.15)$$

where k is the hub-tip ratio, and R_A is the area ratio

$$R_A = \frac{A_2 \cos \alpha_2}{A_1 \cos \alpha_1} \qquad (8.16)$$

Typically, the flow angle α_1 will be zero or slightly negative, i.e., opposite in sense to the blade velocity. The flow angle α_2, however, will always be positive.

The rotor secondary-flow loss coefficient is determined from (8.14) and (8.15) together with the following equations:

$$Y_{sR} = \lambda_s \left(\frac{c_L}{s/c} \right)^2 \frac{\cos^2 \beta_3}{\cos^3 \beta_m} \qquad (8.17)$$

$$c_L = \frac{2s}{c}(\tan \beta_2 - \tan \beta_3) \cos \beta_m \qquad (8.18)$$

and

$$R_A = \frac{A_3 \cos \beta_3}{A_2 \cos \beta_2} \qquad (8.19)$$

Because β_3 is typically negative, the parenthetical part of (8.18) will usually represent a sum.

The tip-clearance loss coefficient needed in (8.9) is found from

$$Y_{tR} = \frac{0.5c_T}{h} \left(\frac{c_L}{s/c}\right)^2 \frac{\cos^2 \beta_3}{\cos^3 \beta_m} \tag{8.20}$$

In the above calculations it is necessary to calculate α_m and β_m. If the velocity diagram is as shown in Figure 8.3, which is typical for reaction turbines, β_2 is taken as positive and β_3 as negative, since the latter angle has a sense opposite to that of the blade velocity thus,

$$W_{um} = \frac{W_{u2} + W_{u3}}{2} \tag{8.21}$$

and

$$\tan \beta_m = \frac{\tan \beta_2 + \tan \beta_3}{2} \tag{8.22}$$

Since β_3 is negative, both W_{u3} and $\tan \beta_3$ are negative, and the numerators of (8.21) and (8.22) represent differences. Similarly, the mean absolute gas angle for the stator is given by

$$\tan \alpha_m = \frac{\tan \alpha_1 + \tan \alpha_2}{2} \tag{8.23}$$

The loss coefficients defined by (8.9) and (8.10) are related to the total pressures through

$$Y_S = \frac{P_{o1} - P_{o2}}{P_{o2} - P_2} \tag{8.24}$$

for the stator, and as

$$Y_R = \frac{P_{o2R} - P_{o3R}}{P_{o3R} - P_3} \tag{8.25}$$

for the rotor. These are useful in determining pressures P_2 and P_3, as well as in the calculation of stage efficiency η_s.

We first modify (8.24) to read

$$Y_S = \frac{P_{o1}/P_2 - P_{o2}/P_2}{P_{o2}/P_2 - 1} \tag{8.26}$$

The ratio P_{o2}/P_2 is easily determined from the basic isentropic relation

$$\frac{P_{o2}}{P_2} = \left[1 + \frac{(\gamma - 1)V_2^2}{2\gamma RT_2} \right]^{\gamma/(\gamma - 1)} \tag{8.27}$$

where T_2 is obtained from the energy equation

$$\frac{T_{o2}}{T_2} = 1 + \frac{(\gamma - 1)V_2^2}{2\gamma RT_2} \tag{8.28}$$

The stagnation temperature is constant in the nozzle, so that $T_{o2} = T_{o1}$, and all inlet conditions are known at the outset. Using the experimental value of Y_S, (8.26) and (8.27) are solved for the ratio P_{o1}/P_2, and P_2 is found from the known value of P_{o1}. The total pressure P_{o2} is then calculated from the ratio determined from (8.27).

The pressure P_3 is found in a similar manner using (8.25). Since the rotor is moving, the relative total pressures P_{o2R} and P_{o3R} are used. They are calculated from

$$\frac{P_{o2R}}{P_2} = \left[1 + \frac{(\gamma - 1)W_2^2}{2\gamma RT_2} \right]^{\gamma/(\gamma - 1)} \tag{8.29}$$

and

$$\frac{P_{o3R}}{P_3} = \left[1 + \frac{(\gamma - 1)W_3^2}{2\gamma RT_3} \right]^{\gamma/(\gamma - 1)} \tag{8.30}$$

where T_3 is found from the energy relation

$$\frac{T_{o3}}{T_3} = 1 + \frac{(\gamma - 1)V_3^2}{2\gamma RT_3} \tag{8.31}$$

using T_{o3} determined from the energy transfer, i.e.,

$$T_{o3} = T_{o1} - \frac{E}{C_p} \tag{8.32}$$

We must use the empirical value of Y_R in (8.25) to calculate P_{o2R}/P_3. Since P_{o2R} is known from (8.29), we can calculate P_3 directly.

Finally, to obtain the stage efficiency η_s using (8.5) and (8.6), we must calculate a value of $T_{o3'}$. Referring to Figure 8.6, it is evident that P_{o3}/P_3 is equal to $P_{o3'}/P_{3'}$, and thus that $V_3^2/T_{3'}$ is equal to V_3^2/T_3. Therefore, we write

$$T_{o3'} = T_{3'} \left[1 + \frac{(\gamma - 1)V_3^2}{2\gamma RT_3} \right] \tag{8.33}$$

where

$$\frac{T_{3'}}{T_{o1}} = \left(\frac{P_3}{P_{o1}} \right)^{(\gamma - 1)/\gamma} \tag{8.34}$$

is used to determine $T_{3'}$. Thus the stage efficiency is easily determined from the empirical loss coefficients Y_S and Y_R.

8.3 Design

As indicated earlier in Chapter 7, a thermodynamic analysis allows the choice of a suitable cycle pressure ratio R_p (see Figure 7.2). Metallurgical considerations govern the choice of the turbine inlet temperature T_{in}. Using an assumed turbine efficiency η_t, the turbine work W_t for the cycle is computed using the relation

$$W_t = c_p \eta_t T_{in} \left[1 - R_p^{-(\gamma -)/\gamma} \right] \tag{8.35}$$

The stage work, or energy transfer E, is related to the turbine work through

$$E = \frac{W_t}{n_{st}} \tag{8.36}$$

where n_{st} is the number of stages. The mass flow rate $\dot{m}$ is related to the stage power through

$$\dot{m} = \frac{P}{E} \tag{8.37}$$

We have already seen how the designer can construct a stage velocity diagram from a knowledge of $\dot{m}$, P, and N. Information from the diagram can be used to determine n_{st} from a formula derived by Vincent (1950), viz.,

$$n_{st} = \frac{W_t \sin^2 \alpha}{V_a^2 (1 - k_B^2 \sin^2 \alpha)} \tag{8.38}$$

where V_a is the assumed axial velocity, α is the nozzle angle, and k_B is a blade friction coefficient, typically having a value between 0.9 and 0.95. The stage formula (8.38) was derived by assuming that all stage velocity diagrams are symmetrical (Figure 8.4) and have the same nozzle angle α and axial velocity V_a. The final design may utilize velocity diagrams different from this, but this assumption is useful in the early stages of the design process.

The velocity diagram is constructed for each of the n_{st} stages in the manner previously discussed in connection with Figure 8.5. The pressures P_2 and P_3 and temperatures T_2 and T_3 are calculated as previously discussed, for the mean diameter. From these values density, flow area, and blade height are calculated at sections 1, 2, and 3.

An alternative method for construction of an optimal velocity diagram is recommended by Vavra (1960). Table 8.1 presents selected results from Vavra's work to obtain the optimum velocity diagram for a chosen degree of reaction. The ideal degree of reaction is used and is defined by

$$R_i = \frac{h_{2'} - h_{3'}}{h_{o1} - h_{3'}} \tag{8.39}$$

Table 8.1 Optimum Gas Turbine Parameters

Parameter	α_2 (degrees)	R_i 0.0	0.25	0.50	0.75
	75	0.462	0.562	0.737	0.837
U/c_o	71	0.437	0.537	0.712	0.837
	66	0.412	0.537	0.712	0.837
	75	0.764	0.797	0.853	0.875
η_{ts}	71	0.731	0.776	0.833	0.863
	66	0.676	0.739	0.804	0.844
	75	0.531	0.378	0.235	0.146
V_a/U	71	0.706	0.498	0.306	0.184
	66	0.936	0.622	0.383	0.230

Source: Vavra (1960).

The values of U/c_o, V_a and η_{ts} presented in Table 8.1 are those corresponding to the highest stage efficiency for the conditions imposed; thus, for the selected R_i and α_2, η_{ts} is the maximum possible.

The spouting velocity c_o is determined from the assumed inlet temperature T_{o1} and inlet and exhaust pressures, p_{o1} and p_3. The mean blade velocity is calculated from the optimum speed ratio U/c_o, or the reverse process may be used, i.e, one in which the blade speed is chosen and c_o and the related pressures and temperature are determined from the optimum speed ratio. For example, Jennings and Rogers (1953) recommend maximum blade tip speeds of 1200 fps, which provides a basis for the calculation of spouting velocity.

Since the stage efficiency can be written as

$$\eta_{ts} = \frac{2U\Delta V_u}{c_o^2}$$

(8.40)

ΔV_u can be calculated from (8.40), and V_{u3} is found from ΔV_u using

$$V_{u3} = V_a \tan \alpha_2 - \Delta V_u$$

(8.41)

Table 8.2 Design Values of Blade Loading Coefficient

			R		
φ	0	0.1	0.3	0.5	0.7
0.5	–	2.0	2.2	4.0	4.4
0.6	2.5	2.6	3.1	4.8	4.5
0.7	3.0	3.2	3.8	4.8	4.5
0.8	3.4	3.6	4.4	4.6	4.4
0.9	3.8	4.1	4.5	4.0	3.6
1.0	4.2	4.4	4.0	3.0	3.0

Source: Wilson (1987)

Thus, the fundamental elements of the velocity diagram are determinable by the Vavra method. The method can be used to obtain a reliable preliminary design of a gas turbine stage in a short time.

Several parameters of the design can be checked to be sure they are in acceptable ranges. The blade loading coefficient ψ_B defined as

$$\psi_B = \frac{2E}{U^2} \tag{8.42}$$

should lie between 1.5 and 5 (Table 8.2). Recommended values of flow angle α_2 are presented in Table 8.3. The flow coefficient φ, defined as

Table 8.3 Design Values of α_2 (Degrees)

			R		
φ	0	0.1	0.3	0.5	0.7
0.5	–	71	69	72	70
0.6	70	70	68	71	68
0.7	68	68	67	68	64
0.8	67	67	65	65	60
0.9	65	65	63	60	55
1.0	64	64	60	–	–

Source: Wilson (1987)

$$\varphi = \frac{V_a}{U} \tag{8.43}$$

would be in the range 0.2 to 1.0. The degree of reaction R, defined as

$$R = \frac{T_2 - T_3}{T_1 - T_3} \tag{8.44}$$

should have a value between 0 and 0.7 at the mean diameter. For the best efficiency, Wilson (1987) recommends using $R = 0.5$ and keeping the hub-tip ratio k between 0.6 and 0.87. The Mach numbers M_1 and M_3, based on V_1 and V_3, respectively, should be subsonic and as low as possible. The nozzle exit Mach number M_2, based on V_2 should be subsonic or slightly supersonic, but not greater than 1.2. The corresponding relative Mach number M_{R2}, based on W_2, should be less than the critical Mach number, which would normally lie in the range 0.7 to 0.9. Adjustments to the velocity diagram can be made, if necessary, to satisfy the above conditions.

The free-vortex condition can be utilized to obtain velocity diagrams at the root and tip of the blade. This is frequently done, since it yields a constant value of energy transfer and of axial velocity, as has been pointed out previously. Twisting of blades, which is required to achieve the free-vortex condition, is often avoided to reduce manufacturing expense. If a condition other than the free-vortex condition is assumed, then the basic equations must be utilized to determine the variation of V_a, i.e., to find V_a at the root and tip of the blade. Then the velocity triangles at root, mean, and tip locations can be constructed, and the energy transfer and mass flow rate can be determined.

The blade angles and dimensions to produce the velocity triangles must be sought. The stator gas angles α_1 and α_2 differ from the stator blade angles α_{B1} and α_{B2}, as shown in Figure 8.9. These differences in angles may be expressed as incidence i, defined by

$$i = \alpha_1 - \alpha_{B1} \tag{8.45}$$

and by deviation δ_B, defined by

$$\delta_B = \alpha_{B2} - \alpha_2 \tag{8.46}$$

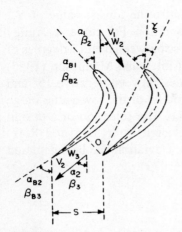

Figure 8.9 Blade angles and gas angles.

For the moving blades we define incidence by

$$i = \beta_2 - \beta_{B2} \tag{8.47}$$

and deviation by

$$\delta_B = \beta_{B3} - \beta_3 \tag{8.48}$$

The profile loss coefficient Y_m, obtained from a cascade test, is usually plotted as a function of incidence, as illustrated in Figure 8.10. Curve A represents data from impulse blading, while curve B depicts reaction-

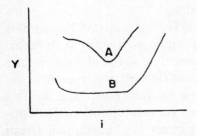

Figure 8.10 Blade loss coefficient from cascade tests.

blading results. The incidence corresponding to the lowest value of Y_m would generally be chosen for the design condition. For curve A, Ainley and Mathieson (1955) found the minimum loss between 5 and 7 degrees of positive incidence. In the case of curve B, Ainley and Mathieson (1955) found that minimum loss occurs for incidences between negative 15° and positive 15°. A design incidence near the middle of the low-value range should result in good off-design performance. The selection of a design incidence allows us to calculate the blade inlet angle from (8.45) and (8.47) using gas angles from the velocity diagrams. Deviation can be calculated from an empirical relation given by Horlock (1973):

$$\delta_B = m\theta \left(\frac{s}{c}\right)^{\frac{1}{2}} \tag{8.49}$$

where the camber θ is estimated by

$$\theta = |\alpha_{B1} - \alpha_{B2}| \tag{8.50}$$

for the stator, and by

$$\theta = |\beta_{B2} - \beta_{B3}| \tag{8.51}$$

for the rotor. The constant m depends on stagger angle γ_s (Figure 8.9). For the circular-arc camber line, m varies from 0.21 at $\gamma_S = 0°$ to 0.17 at $\gamma_S = 60°$. For the parabolic-arc camber line, m is 0.12 at 0° stagger angle and is 0.06 at 60° stagger angle. Thus, knowing i and δ_B we are able to determine the blade angles for stator and rotor.

Although current research, such as that of Nicoud et al. (1991), promises computational design of the blade profile to suit prescribed velocity distributions, blades are often laid out in the form of an airfoil along a circular arc or parabolic camber line using a chord c of from 80 to 90 percent of the blade length. A typical (maximum) thickness-to-chord ratio is 0.2, and this maximum thickness is located at the 40 percent chord position. The optimum spacing between blades at the mean diameter may be determined from the relations given by Shepherd (1956), i.e.,

$$\frac{c}{s} = 2.5 \cos^2 \alpha_2 \frac{\tan \alpha_2 - \tan \alpha_1}{\cos \alpha_m} \tag{8.52}$$

for the stator, and from

$$\frac{c}{s} = 2.5 \cos^2 \beta_3 \frac{\tan \beta_2 - \tan \beta_3}{|\cos \beta_m|} \tag{8.53}$$

for the rotor. Generally, the value of c/s will lie between 1.0 and 1.8.

The aerodynamic design is thus complete. However, calculation of centrifugal, bending, and disk stresses must also be carried out to ensure safe operation with available materials. Stress calculations are treated by Jennings and Rogers (1953), Cohen et al. (1987), and Mattingly et al. (1987).

8.4 Examples

Example Problem 8.1

A velocity diagram is established with U = 1000 fps, $\alpha_2 = 60°$, $\alpha_3 = -10°$, and $\varphi = 0.8$. The inlet total temperature is 1000°R, and the stator area ratio $A_2/A_1 = 1.3$. Assuming that the hub-tip ratio k = 0.8, tip clearance $c_T/h = 0.02$, thickness ratio t/c = 0.2, and $V_{a2} = V_{a3}$, find Y_R and Y_S.

Solution: Solve for flow angles and velocities from the given velocity diagram using trigonometric relations. The resulting values are the following: $V_a = 800$ fps; $V_{u2} = 1386$ fps; $V_{u3} = -141$ fps; $V_2 = 1600$ fps; $V_3 = 812$ fps; $\beta_2 = 25.76°$; $\beta_3 = -54.96°$; $\beta_m = -25.27°$; $\alpha_m = 40.89°$; ($\alpha_1 = 0°$). From the Euler equation (7.6),

$$E = 1000[1385.6 - (-141)] = 1,526,600 \text{ ft}^2/\text{s}^2$$

Solve (8.52) for solidity. For the stator:

$$\frac{c}{s} = \frac{2.5 \cos^2 60 \tan 60}{\cos 40.89} = 1.432 \qquad \frac{s}{c} = 0.698$$

Enter Figure 8.8 to find profile loss coefficient.

$$Y_{ms} = 0.028$$

Use (8.13) to find the lift coefficient.

$$\frac{c_L}{s/c} = 2 \tan 60 \cos 40.89 = 2.619$$

Solve for the secondary flow loss parameter using (8.14), (8.15), and (8.16).

$$R_A = \frac{1.3 \cos 60}{\cos 0} = 0.65$$

$$x = \frac{(0.65)^2}{1 + 0.8} = 0.2347$$

$$\lambda_s = 0.0996(0.2347)^{2.1854} + 0.0056 = 0.0098$$

Use (8.12) to find the secondary flow loss coefficient.

$$Y_{ss} = 0.0098(2.619)^2 \left(\frac{\cos^2 60}{\cos^3 40.89} \right) = 0.0389$$

Finally, the stator loss coefficient is

$$Y_s = 0.028 + 0.0389 = 0.0669$$

Solve (8.53) for solidity. For the rotor:

$$\frac{c}{s} = \frac{2.5 \cos^2 54.96[\tan 25.73 - \tan(-54.96)]}{\cos 25.27} = 1.74$$

$$\frac{s}{c} = 0.575$$

Use Figure 8.7 to get $Y_{mI} = 0.092$. Use Figure 8.8 to get $Y_{mo} = 0.034$. Substitute these into (8.11) to get

$$Y_{mR} = 0.034 + \left(\frac{25.73}{54.96}\right)^2 (0.092 - 0.034) = 0.0467$$

Use (8.18) to find the lift coefficient.

$$\frac{c_L}{s/c} = 2[\tan 25.73 - \tan(-54.96)] \cos(-25.27) = 3.45$$

Calculate the secondary flow loss parameter as above.

$$R_A = \frac{1.38 \cos(-54.96)}{\cos 25.73} = 0.8827$$

where the area ratio is estimated from the continuity equation. $A_2 \rho_2 V_{a2} = A_3 \rho_3 V_{a3}$; noting that V_a is constant in the rotor and taking the expansion process to be isentropic, we have

$$\frac{A_3}{A_2} = \left(\frac{T_2}{T_3}\right)^{1/(\gamma - 1)} = \left(\frac{786.7}{690.6}\right)^{2.5} = 1.385$$

where temperatures are found from (7.23), (7.24), and (7.51); thus,

$$T_2 = 1000 - \frac{(1600)^2}{12,000} = 786.7°R$$

$$T_{o3} = 1000 - \frac{1,526,600}{6000} = 745.6°R$$

$$T_3 = 745.6 - \frac{(812.3)^2}{12,000} = 690.6°R$$

$$x = \frac{(0.8827)^2}{1 + 0.8} = 0.4329$$

$$\lambda_s = 0.0996(0.4329)^{2.1854} + 0.0056 = 0.02158$$

Substitute the above into (8.17) to get

$$Y_{sR} = 0.02158(3.45)^2 \frac{\cos^2 (-54.96)}{\cos^3 (-25.27)} = 0.1145$$

Solve for the tip clearance loss coefficient using (8.20).

$$Y_{tR} = \frac{0.5(0.02)(3.45)^2 \cos^2 (-54.96)}{\cos^3 (-25.27)} = 0.0531$$

Sum the losses using (8.9) to obtain

$$Y_R = 0.0467 + 0.1145 + 0.0531 = 0.2143$$

Example Problem 8.2

Use the Vavra method to determine relative flow angles and hub and tip radii for a single-stage reaction air turbine with $R_i = 0.5$ and $\alpha_2 = 75$ degrees. The turbine should deliver 100 hp at 50,000 rpm with $p_{o1} = 3$ atmospheres, $T_{o1} = 560°R$ and $p_3 = 1$ atmosphere.
Solution: From Table 8.1, $U/c_o = 0.737$, $\eta_{ts} = 0.853$, and $V_a/U = 0.235$.
Calculate c_o using (7.40). $c_o = 1346$ fps.
From the speed ratio, $U = 0.737(1346) = 992$ fps. From the flow coefficient, $V_a = 0.235(992) = 233.1$ fps. Using equation (8.40),

$$\Delta V_u = \frac{0.853(1346)^2}{2(992)} = 778.9 \text{ fps}$$

From (7.6), $E = 992(778.9) = 772,697$ ft^2/s^2. Using the velocity triangles and trigonometric relations, we find that $V_{u2} = 870$ fps; $V_{u3} = 91.1$ fps; $W_{u2} = -122$ fps; $V_3 = 263.1$ fps; $\beta_3 = -75.49°$; and $\beta_2 = -27.6°$. From (8.37),

$$\dot{m} = \frac{100(550)(32.174)}{772697} = 2.29 \text{ lb/s}$$

Using (7.24) and (7.51), calculate the exhaust temperature.

$$T_{o3} = 560 - \frac{772,697}{6000} = 431.2°R$$

$$T_3 = 431.2 - \frac{(263.1)^2}{12,000} = 425.4°R$$

Using the perfect gas equation of state, calculate exhaust gas density,

$$\rho_3 = \frac{14.7(144)}{53.3(425.4)} = 0.0933 \text{ lb/cu ft}$$

Using the mass flow equation, solve for A_3.

$$A_3 = \frac{2.29(144)}{0.0933(233.1)} = 15.15 \text{ sq in}$$

Calculate the mean blade radius.

$$r_m = \frac{U}{N} = \frac{992(12)}{5235.98} = 2.2735 \text{ in}$$

Solve for hub and tip radii from the area equation,

$$A_3 = \pi(r_t^2 - r_H^2)$$

and the rms equation,

$$2r_m^2 = r_t^2 + r_H^2$$

The results are $r_t = 2.753$ in and $r_H = 1.66$ in.

References

Ainley, D. G., and G. C. R. Mathieson. 1955. *An Examination of Flow and Pressure Losses in Blade Rows of Axial-Flow Turbines*. ARC R&M No. 2891. HMSO, London.

Cohen, H., G. F. C. Rogers, and H. I. H. Saravanamuttoo. 1987. *Gas Turbine Theory*. Longman, London.

Horlock, J. H. 1973. *Axial Flow Turbines*. Krieger, Huntington, New York.

Jennings, B. H., and W. L. Rogers. 1953. *Gas Turbine Analysis and Practice*. McGraw-Hill, New York.

Kacker, S. C., and U. Okapuu. 1982. A Mean Line Prediction Method for Axial Flow Turbine Efficiency. *Journal of Engineering for Power. 104:* 111-119.

Mattingly, J. D., W. H. Heiser and D. H. Daley. 1987. *Aircraft Engine Design*. AIAA, New York.

Nicoud, D., LeBloa, C. and O. P. Jacquotte. 1991. A Finite Element Inverse Method for the Design of Turbomachinery Blades. ASME Paper 91-GT-80.

Shepherd, D. G. 1956. *Principles of Turbomachinery*. Macmillan, New York.

Vavra, M. H. 1960. *Aerothermodynamics and Flow in Turbomachines*. John Wiley & Sons, New York.

Vincent, E. T. 1950. *The Theory and Design of Gas Turbines and Jet Engines*. McGraw-Hill, New York.

Wilson, D. G. 1987. New Guidelines for the Preliminary Design and Performance Prediction of Axial-Flow Turbines. *Proc. Inst. Mech. Engrs., 201:* 279-290.

Problems

8.1   Estimate the number of stages needed for a 50 percent reaction turbine operating at standard sea level in a basic open-cycle gas-turbine plant. Turbine inlet temperature is to be 1660°R. Assume cold-air properties throughout, nozzle angle = 20°, axial component of velocity = 420 ft/s, minimum leaving kinetic energy, η_t = 0.90, and blade friction coefficient K_B = 0.90. Shaft speed is 10,000 rpm and shaft power is 9000 hp. Cycle pressure ratio is 8.3.

8.2   Draw a mean velocity diagram for the turbine described in Problem 8.1

8.3   Estimate hub and tip diameters for the last stage of the turbine of Problem 8.1.

8.4   A gas-turbine stage is designed for which T_{o1} = 1100 K, P_{o1} = 4 bar, V_2 = 519 m/s, U = 340 m/s, V_{a2} = V_{a3} = 272 m/s, α_3 = −10°, T_{o3} = 955°K, and β_2 = 20.5° (Figure 8.6). The loss coefficients are Y_S = 0.0688 and Y_R = 0.152. Determine stage efficiency (total-to-total).

8.5   An impulse stage of a gas turbine receives an axially directed stream
      of air at $T_1 = 2080°F$ and $V_1 = 956$ ft/s. At the nozzle exit $T_2 = 1678°F$
      and $M_2 = 1.313$. $A_2/A_1 = 1.5$. The axial component of V_2 equals the
      velocity V_3, which is axially directed. Assume the average ratio of
      specific heats is 1.3. Find:
      a) exhaust velocity
      b) nozzle exit velocity
      c) blade velocity
      d) energy transfer
8.6   The 50 percent reaction stage of a small, single-stage gas turbine has
      an annular flow area of 15.2 square inches and a mean blade radius at
      the rotor exit of 2.4 inches. Exhaust gas temperature T_3 is 426°R and
      exhaust pressure P_3 is 14.7 psia. The speed of the turbine rotor is
      50,000 rpm. $V_{a2} = V_{a3} = 235$ ft/s and $V_{u2} = 1200$ fps. Assume air
      properties with the ratio of specific heats equal to 1.4. Find:
      a) nozzle angle
      b) mean blade speed
      c) mass rate of flow of gas
      d) energy transfer
      e) turbine horsepower
8.7   A single-stage reaction turbine produces 110 horsepower at a turbine
      rotor speed of 45,000 rpm. The relative gas angle is –27 degrees at the
      rotor inlet and –75 degrees at the rotor exit. The stage flow coefficient
      is 0.235, the spouting velocity is 1305 ft/s and the stage pressure ratio
      P_{o1}/P_3 is 2.2. The mean radius of the rotor blade is 2.5 inches. The
      exhaust pressure is 1 atmosphere. Find:
      a) turbine inlet temperature T_{o1}
      b) turbine exhaust temperature T_{o3}
      c) total-to-total efficiency
      d) tip radius at exhaust plane
      e) hub radius at exhaust plane
8.8   An axial-flow gas turbine stage has a constant axial velocity at all
      stations. Flow at the entry and exit of the stage is purely axial. The
      flow coefficient is 0.6 and the nozzle angle is 68.2 degrees. Find:
      a) blade loading coefficient
      b) the relative gas angles
      c) the degree of reaction
8.9   A single-stage axial-flow air turbine produces 110 hp at a speed of
      50,000 rpm. At the nozzle inlet the total pressure is 2.2 atmospheres

and the total temperature is 300°F. At the rotor exhaust the static pressure is one atmosphere. The mean blade speed is 1001 fps. The relative gas angles are −27° and −75.4° at the rotor inlet and exit, respectively, and the flow coefficient is 0.235. Calculate:
- a) spouting velocity
- b) mean blade radius
- c) energy transfer
- d) required mass flow
- e) total-to-total efficiency

8.10 Air enters a single-stage axial-flow turbine at 560°R (total temperature) and leaves at 431°R (total temperature). The relative velocity at the rotor exit is 926 fps. Other data are (refer to Fig. 8.4): $T_2 = 492$°R, $T_{2'} = 48$°R, $T_3 = 426$°R, $T_{3'} = 409$°R, $T_{3''} = 416$°R. Find:
- a) nozzle loss coefficient defined by (7.52)
- b) rotor loss coefficient defined by (7.55)

8.11 A single-stage, axial-flow air turbine will produce 110 horsepower when the supply air enters at a total pressure of 3 atmospheres and a total temperature of 540°R and exhausts at a static pressure of 1 atmosphere. The turbine speed is 50,000 rpm. Use Table 8.1 with $\alpha_2 = 75$° and $R_i = 0.5$ to construct the velocity triangle. Assume $V_{a2} = V_{a3}$ and $V_1 = V_3$. Without using the Ainley-Mathieson charts, find:
- a) nozzle loss coefficient defined by (7.52)
- b) rotor loss coefficient defined by (7.55)
- c) nozzle loss coefficient defined by (8.24)
- d) rotor loss coefficient defined by (8.25)
- e) total-to-total efficiency

8.12 A single-stage, axial-flow air turbine produces 120 hp at a speed of 50,000 rpm. At the nozzle inlet the total pressure is 2.36 atmospheres, and the total temperature is 300°F. At the rotor exhaust the static pressure is 1 atmosphere. The mean blade speed is 1000 fps. The relative gas angle at the rotor inlet is −27°, and at the rotor exit it is −75.4°. The axial component of velocity is constant through the rotor, and the flow coefficient is 0.245. Find:
- a) spouting velocity
- b) energy transfer
- c) required mass flow

8.13 Find the horsepower produced by an axial-flow air turbine stage having a symmetrical velocity diagram in which $V_{u2} = 870$ fps. The static temperature at the exhaust is 426°R, and the static exhaust pressure is 14.7 psia. The annulus area at the rotor exit is 15.2 sq in, and the mean radius is 2.27 inches. The rotational speed is 5236 radians/sec, and $V_a = 233$ fps and is constant in the rotor.

8.14 An axial-flow impulse air turbine has a mean blade speed of 936 fps. The air flow rate is 75 lb/s. The air enters the stator at a total temperature of 1400°R and exits from the nozzles at a Mach number of 1.2. γ is 1.35. The gas leaves the rotor in an axial direction. The blades are designed for a free-vortex velocity distribution. Find:

     a) V_2
     b) W_2
     c) V_3
     d) α_2
     e) W_3
     f) turbine horsepower

8.15 Use the Vavra method to determine the principal features of a single-stage air turbine that will produce 110 hp at 50,000 rpm. Assume an inlet total pressure of 3 atmospheres and an outlet static pressure of 1 atmosphere. The total temperature at the nozzle outlet is 540°R. Assume $R_i = 0.5$ and $\alpha_2 = 75°$ at the mean radius.

8.16 Use the Ainley-Mathieson method to determine Y_s and Y_R from the data given in Problem 8.4. Further data are: $r_m = 0.216$ m, $A_2/A_1 = 1.33$ and $A_3/A_2 = 1.26$.

Symbols for Chapter 8

A_2	annulus area at rotor inlet
A_3	annulus area at rotor exit
c	chord length = distance from leading edge to trailing edge of blade profile
c/s	solidity
c_L	lift coefficient
c_o	spouting velocity
c_p	specific heat of gas at constant pressure
c_T	tip clearance
c_v	specific heat of gas at constant volume
D_H	hub diameter
D_m	diameter at mean position between hub and tip of blade
D_r	root diameter = D_H

D_s	specific diameter
D_t	tip diameter
$\dot{E}$	energy transfer $= U\Delta V_u = U\Delta W_u$
E	rate of energy transfer $= P$
h	blade height $= r_t - r_H$
h/c	aspect ratio
h_{o1}	total enthalpy of gas at stator inlet
h_{o2}	total enthalpy of gas at rotor inlet
h_{o3}	total enthalpy of gas at rotor exit
$h_{2'}$	static enthalpy after isentropic expansion to p_2
$h_{3'}$	static enthalpy after isentropic expansion to p_3
i	incidence
k	hub-tip ratio
$\dot{m}$	mass flow rate of gas through annulus
N	rotor speed
N_s	specific speed
n_{st}	number of stages in a multistage turbine
P	power produced by turbine stage
p_{o1}	total pressure at stator inlet
p_{o2}	total pressure at stator exit or at rotor inlet
p_{o3}	total pressure at rotor exit
p_{o2R}	relative total pressure at rotor inlet
p_{o3R}	relative total pressure at rotor exit
p_2	static pressure at stator exit
p_3	static pressure at rotor exit
R	degree of reaction
R_A	ratio of exit normal flow area to inlet normal flow area
R_i	ideal degree of reaction
r_H	hub radius or radius at root of blade
r_m	mean blade radius $= D_m/2 = (r_t^2 + r_H^2)^{1/2}$
R_p	cycle pressure ratio
r_t	radius at blade tip
s	spacing between adjacent blades
s/c	pitch-chord ratio
t	maximum blade profile thickness
T_{in}	turbine inlet temperature for multistage turbine
T_{o1}	total temperature at stator inlet
T_{o2}	total temperature at stator outlet or rotor inlet
T_{o3}	total temperature at rotor outlet

T_2	static temperature at rotor inlet
T_3	static temperature at rotor exit
$T_{2'}$	static temperature at state 2'
$T_{3'}$	static temperature at state 3'
U	blade speed at mean radius r_m of blade
U_t	blade speed at tip radius r_t of blade
U_H	blade speed at hub radius r_H of blade
V	absolute velocity of fluid
V_1	absolute velocity of fluid entering stator
V_2	absolute velocity of fluid leaving stator or entering rotor
V_3	absolute velocity of gas leaving rotor
$V_{3'}$	absolute velocity of gas after expansion to state 3'
V_a	component of V in axial direction
V_r	component of V in radial direction
V_u	component of V in tangential direction
V_{u2}	tangential component of V_2
V_{u3}	tangential component of V_3
ΔV_u	$V_{u2} - V_{u3}$
W	velocity of fluid relative to moving blade
W_2	relative velocity entering rotor
W_3	relative velocity leaving rotor
W_u	tangential component of W
ΔW_u	$W_{u2} - W_{u3} = \Delta V_u$
W_{u2}	tangential component of W_2
W_{u3}	tangential component of W_3
x	$R_A^2/(1 + k)$
Y_R	rotor total loss coefficient
Y_s	stator total loss coefficient
Y_{mR}	rotor profile loss coefficient
Y_{ms}	stator profile loss coefficient
Y_{sR}	rotor secondary flow loss coefficient
Y_{ss}	stator secondary flow loss coefficient
Y_{tR}	rotor tip-clearance loss coefficient
α_1	angle between V_1 and V_a
α_2	angle between V_2 and V_a
α_3	angle between V_3 and V_a
α_m	mean absolute gas angle
α_{B1}	angle between leading edge tangent to camber line of a stator vane profile and the axial direction

α_{B2} angle between trailing edge tangent to camber line of a stator vane profile and the axial direction

β_2 angle between W_2 and V_a

β_3 angle between W_3 and V_a

β_{B2} angle between leading edge tangent to camber line of a rotor blade profile and the axial direction

β_{B3} angle between trailing edge tangent to camber line of rotor blade profile and the axial direction

β_m mean relative gas angle

γ ratio of specific heats $= c_p/c_v$

γ_s stagger angle

δ_B deviation

η_s stage efficiency $= \eta_{tt}$ or η_{ts}

η_{tt} total-to-total efficiency

η_{ts} total-to-static efficiency

η_t overall efficiency for multistage turbines

θ camber angle

λ_s secondary flow loss parameter

ρ fluid density

σ solidity at mean radius of blade $= c/s$

φ flow coefficient $= V_a/U$

ψ_B blade loading coefficient

9 Steam Turbines

9.1 Introduction

Steam turbines, like gas turbines, are predominantly axial-flow units. They are used extensively in power plants to drive electric generators, as are gas turbines, but are usually much larger than gas turbines. In addition, the large units typically use much higher pressures in the first stages and lower pressures in their later stages than do the large gas turbines.

Steam-turbine calculations are different than gas-turbine calculations in that tables of steam properties are substituted for simple gas relations. The basic features of the axial-flow turbine are the same for both steam and gas turbines; however, the steam can contain droplets of liquid water, making the flow a two-phase flow, and thus a further complication is added. The simplest ideal cycle in which the steam turbine performs a function is the Rankine cycle. This cycle is pictured on the temperature-entropy diagram of Figure 9.1. The line drawn from point 1 to point 2 represents the ideal expansion of steam in a turbine from a superheated state to a wet state. The actual process, i.e., that with an entropy increase, is shown as a dashed line 1–2'. The wet steam is exhausted from the turbine into a condenser where it is condensed to a saturated liquid at point 3. A pump and boiler then act upon the water to raise its pressure in process 3–4 and heat it in process 4–1. The boiler, or steam generator, delivers the superheated steam to the first stage of the turbine in state 1.

The thermal efficiency of the power-plant cycle is calculated by dividing the net work, i.e., the turbine work minus the pump work, by the heat supplied in the boiler. Although the cycles used in modern power plants

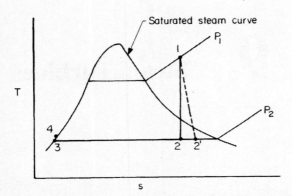

Figure 9.1 Temperature-entropy diagram for a steam power cycle.

involve reheating the steam at one or more points during the expansion and other complexities, the turbine designer can enhance overall cycle efficiency by simply increasing the turbine work realized during the expansion 1–2'. As in previous chapters, the individual stage design will be emphasized in the present chapter. After the decisions regarding the cycle thermodynamics are made, and the inlet and exhaust conditions are thus established, the turbine designer must divide the expansion process into smaller stage pressure or enthalpy drops and then proceed to design individual stages. Both impulse and reaction axial-flow stages will be considered, since the former does find application in the early stages of large reaction turbines, and the impulse stage is used in small turbines. The Ljungstrom turbine, which utilizes a radial flow of steam, will not be treated here. The latter is analyzed by Shepherd (1956).

9.2   Impulse Turbines

If the turbine stage is to be an impulse stage, the entire pressure drop must occur in the nozzles. The purpose of the moving blade is to reduce the kinetic energy of the steam and transfer this energy to work done on the moving blades. The resulting energy transfer may be evaluated using (2.19):

$$E = \frac{V_2^2 - V_3^2}{2} \tag{9.1}$$

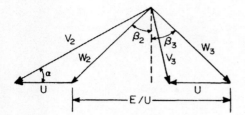

Figure 9.2 Velocity diagram for an impulse turbine.

where the subscripts 2 and 3 refer to rotor inlet and exit, respectively. Typically, the velocity diagram is that shown in Figure 9.2. Note that the relative velocity W_2 is equal in magnitude to W_3. Thus in the ideal impulse turbine the relative velocity changes direction from β_2 to β_3, but the magnitude holds constant. However, the absolute velocity leaving the rotor is much reduced from the velocity V_2 exiting from the nozzles.

Since the boundary layers which form on the blade surfaces actually slow the steam in the passage between the blades, velocity coefficients K_S and K_R are usually defined for the stator and rotor, respectively, by the following relations:

$$K_S = \frac{V_2}{V_{2'}}$$
(9.2)

and

$$K_R = \frac{W_3}{W_2}$$
(9.3)

where the thermodynamic states 1, 2, 2', and 3 are indicated in Figure 9.3. These coefficients can be estimated from values given in Table 9.1. The values in Table 9.1 were calculated from empirical data presented by Horlock (1973), although the small effect of the Reynolds number is neglected. The quantity $\Delta\beta$ denotes the deflection of fluid by the stator or rotor blades, and H/c_a denotes the blade aspect ratio, i.e., blade height H over axial chord c_a.

The Euler turbine equation (2.16) applied to the steam turbine becomes either

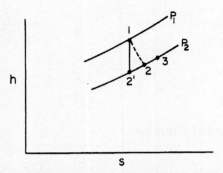

Figure 9.3 Expansion process in an impulse turbine.

$$E = U(V_{u2} - V_{u3}) \tag{9.4}$$

or

$$E = U(W_{u2} - W_{u3}) \tag{9.5}$$

Applying the latter form and expressing components in terms of the steam angles β_2 and β_3 relative to the moving blades, we have

$$E = U(W_2 \sin \beta_2 + W_3 \sin \beta_3) \tag{9.6}$$

Using (9.3) we have

$$E = UW_2 (\sin \beta_2 + K_R \sin \beta_3) \tag{9.7}$$

Table 9.1 Velocity Coefficients K_S and K_R

H/c_a	$\Delta\beta(\deg)$				
	20	40	60	90	100
1	0.96	0.95	0.95	0.94	0.93
3	0.98	0.98	0.97	0.96	0.95
∞	0.99	0.99	0.98	0.97	0.97

Using the relation, derived from Figure 9.2, that

$$W_2 \sin \beta_2 = V_2 \cos \alpha - U \tag{9.8}$$

(9.7) becomes

$$E = U(V_2 \cos \alpha - U)\left(1 + \frac{K_R \sin \beta_3}{\sin \beta_2}\right) \tag{9.9}$$

If we then define a blade efficiency η_B by

$$\eta_B = \frac{2E}{V_2^2} \tag{9.10}$$

which is a reasonable measure of the effectiveness of impulse-turbine blading, we find that

$$\eta_B = \frac{2U}{V_2}\left(\cos \alpha - \frac{U}{V_2}\right)\left(1 + \frac{K_R \sin \beta_3}{\sin \beta_2}\right) \tag{9.11}$$

By differentiation with respect to U/V_2 we are able to find the optimum ratio U/V_2 for maximum blade efficiency:

$$\frac{U}{V_{2opt}} = \frac{\cos \alpha}{2} \tag{9.12}$$

Since the nozzle angle α usually falls between $15°$ and $25°$, we expect the optimum ratio of blade speed to nozzle exit speed to lie between 0.483 and 0.453.

The actual value of η_{Bmax} depends on the values of K_B, β_2, and β_3, but it is usually quite high. It varies, however, from zero at zero blade speed to zero at $U = V_2$, as shown in Figure 9.4. Torque as a function of blade speed is also shown in Figure 9.4. Torque, denoted by T in Figure 9.4, varies in a straight line from its maximum at zero blade speed to zero at $U = V_2$. This is easily shown by noting that torque is power over rotational speed. Thus

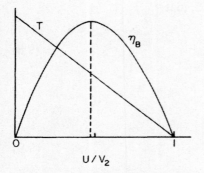

Figure 9.4 Efficiency and torque for an impulse turbine.

$$T = \frac{\dot{E}}{N} = \frac{\dot{m}E}{N}$$

(9.13)

and

$$U = \frac{ND}{2}$$

(9.14)

Substitution of (9.9) and (9.14) into (9.13) yields

$$T = \frac{\dot{m}D}{2}\left(V_2 \cos \alpha - \frac{ND}{2}\right)\left(1 + \frac{K_R \sin \beta_3}{\sin \beta_2}\right)$$

(9.15)

in which T is a linear function of N. The design torque T_D is the torque at maximum efficiency and is obtained by substitution of (9.12) into (9.15). This gives

$$T_D = \frac{\dot{m}D}{4}(V_2 \cos \alpha)\left(1 + \frac{K_R \sin \beta_3}{\sin \beta_2}\right)$$

(9.16)

A velocity-compounded, or Curtis, stage is sometimes used when the blade speed is lower than optimum, and it is desirable to extract more kinetic

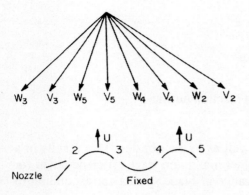

Figure 9.5 Curtis stage.

energy from the steam than is possible with a single row of moving blades. An additional row of moving blades is added to the same turbine wheel, and a fixed row of blades is interposed between the moving rows, as is shown in Figure 9.5. Thus the steam leaves the fixed row having a velocity V_4 only lightly less than V_3 and then passes over the second row of moving blades, leaving with velocity V_5, which ideally is axial. Additional moving rows may also be added if required to reduce the absolute exit velocity to its minimum (axial) value. An analysis of blade efficiency, similar to that carried out above, shows that

$$\frac{U}{V_{2opt}} = \frac{\cos \alpha}{4} \qquad (9.17)$$

for the stage depicted in Figure 9.5 with $K_R = K_S = 1$. Thus (9.17) indicates one-half the blade speed needed for maximum efficiency with only one row of moving blades and thus illustrates that velocity compounding implies lower blade speeds for the same nozzle exit velocity.

9.3 Reaction Turbines

For a large steam turbine, the designer could choose to use pressure-compounding, i.e., use a row of nozzles after each row of moving impulse blades. A more effective procedure, however, is to allow the expansion to

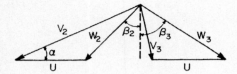

Figure 9.6 Velocity diagram for a reaction turbine.

proceed in the moving blades as well as in the nozzles. This results in a higher optimum blade speed, a higher maximum stage efficiency than for the impulse stage, and a broader range of blade speeds corresponding to high efficiency.

Referring to the velocity diagram in Figure 9.2 again, we see that a reaction turbine diagram would be similar, but would have W_3 considerably longer than W_2. This follows from the steady-flow energy equation (2.9) and the Euler turbine equation (2.18). Eliminating the energy transfer results in the form

$$h_2 + \frac{W_2^2}{2} = h_3 + \frac{W_3^2}{2} \tag{9.18}$$

Thus a pressure drop in the moving blades is accompanied by a loss of enthalpy and a gain in relative kinetic $W_3^2/2$. Figure 9.6 illustrates this increase of relative velocity, and from it we should also observe that the blade would not have the symmetry that would be expected in an impulse blade.

The velocity coefficients from Table 9.1 can still be used in the analysis of this type of turbine if we approximate the enthalpy rise associated with blade friction in the same manner used previously in the nozzles of the impulse turbine. Thus for the stator of the reaction turbine we can write

$$h_{01} = h_2 + \frac{V_2^2}{2} = h_{2'} + \frac{(V_{2'})^2}{2} \tag{9.19}$$

where the states 2 and 2′ are those depicted in Figure 9.7. Using (9.2) to substitute for V_2, we have

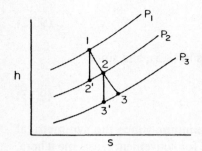

Figure 9.7 Thermodynamic processes in reaction turbines.

$$h_2 - h_{2'} = \frac{(V_{2'})^2 - K_S^2(V_{2'})^2}{2}$$

(9.20)

Since $V_{2'}$ is the ideal velocity resulting from an isentropic expansion, it is easily found, and then (9.20) can be used to find h_2, the actual enthalpy leaving the nozzle.

The same method applied to the rotor, but using relative velocities in place of absolute velocities, leads to

$$h_3 - h_{3'} = \frac{(W_{3'})^2 - K_S^2(W_{3'})^2}{2}$$

(9.21)

Thus, knowing pressures P_1, P_2, and P_3, we can easily determine $V_{2'}$ and $W_{3'}$, and we can then use (9.20) and (9.21) to estimate the actual enthalpies h_2 and h_3. The velocities V_2 and W_3 are also easily obtained from the velocity coefficient.

With the velocities V_2 and W_3 known, a velocity diagram, such as that shown in Figure 9.6, is easily constructed for a given nozzle angle α and blade speed U. The axial and tangential components of V_2 are easily calculated and used to find W_2. Assuming that a prior knowledge of $\Delta\beta$ is required to select K_S, we thus know β_3 and we can determine W_{u3} from W_3, and finally V_{u3} and V_3. Thus all the velocities and angles may be determined using the velocity coefficient K_S to determine the enthalpy rise due to friction.

Energy transfer can be found with the help of Figure 9.6. The Euler turbine equation becomes

$$E = U(V_2 \cos \alpha + W_3 \sin \beta_3 - U) \tag{9.22}$$

Since the stage efficiency can be defined as in (8.5), and since

$$E = h_{01} - h_{03} \tag{9.23}$$

efficiency is a maximum when E is a maximum. Since the symmetrical velocity diagram is commonly used, we will, for convenience, assume it here, i.e., we assume that $V_2 = W_3$ and $V_3 = W_2$. For this case, (9.22) becomes

$$E = U(2V_2 \cos \alpha - U) \tag{9.24}$$

Differentiation with respect to U/V_2 shows that maximum efficiency corresponds to

$$\frac{U}{V_{2opt}} = \cos \alpha \tag{9.25}$$

which is twice the optimum blade speed found for the impulse turbine. As mentioned previously the resulting stage efficiency is higher and the curve flatter than for the impulse turbine, implying better off-design performance as well. The range of speed ratios for high efficiency is roughly $0.7 < U/V_2 < 1.3$.

9.4 Design

The design of steam turbines will proceed along the same lines previously outlined in Chapter 8 for gas turbines. Of course, steam tables or Mollier charts must be applied in lieu of perfect-gas relations for the determination of properties, but otherwise the same methods apply. For a detailed account of steam-turbine construction and design, the reader is referred to a book by Church (1950).

References

Church, E. F. 1950. *Steam Turbines*. McGraw-Hill, New York.

Dietzel, F. 1980. *Dampfturbinen.* Hanser, München.

Elliott, T. C. 1989. *Steam Turbine Fundamentals. Standard Handbook of Power Plant Engineering.* McGraw Hill, New York.

Horlock, J. H. 1973. *Axial Flow Turbines.* Krieger, Huntington, New York.

Shepherd, D. G. 1956. *Principles of Turbomachinery.* Macmillan, New York.

Stodola, A. 1927. *Steam and Gas Turbines.* McGraw-Hill, New York.

Problems

9.1 Construct a velocity diagram for a single-stage impulse turbine having a nozzle angle $\alpha = 20°$, a mean radius $r_m = 10$ in., a blade height H of 2 in., a speed N = 7200 rpm, an aspect ratio $H/C_a = 2$, and equal relative steam angles ($\beta_2 = \beta_3$). At the nozzle exit, the enthalpy $h_2 = 1183.5$ Btu/lb, pressure $p_2 = 14.0$ psia, and $V_2 = 1256.6$ ft/s.

9.2 Determine the power produced by the turbine in Problem 9.1, assuming full admission through all nozzles in the annulus.

9.3 Assuming that turbine power is varied by progressively blocking nozzles, i.e., through the use of partial admission, show that the steam consumption of the turbine in Problem 9.1 is a linear function of turbine power. This straight line is the Willans line. Plot the Willans line using lb/hr units for steam flow and kilowatts for turbine power.

10 Hydraulic Turbines

10.1 Introduction

The oldest form of power-producing turbine is that utilizing the motive power of water. Flowing rivers, streams, and waterfalls have had their energy extracted by vanes or buckets fixed to the circumference of rotating wheels. Although these waterwheels were designed and built as late as the nineteenth century, and some of them were fairly efficient, they eventually gave way to more powerful machines requiring reservoirs.

Modern installations utilize a reservoir of water, which is usually water collected from a flowing river. The level of the reservoir next to the hydraulic power station is maintained at a nearly constant elevation by controlling inflow from other reservoirs further upstream. The water flows into the hydraulic turbine through a large pipe, known as a penstock. It leaves the turbine through a diverging duct, known as a draft tube, and enters a downstream reservoir, known as a tail race. The available head, which ranges from several feet to several thousand feet in existing plants, is the vertical distance between the free surfaces of the water in the reservoir and the tail race. A schematic diagram is shown in Figure 10.1.

The designer is generally presented with an available flow rate Q, based on runoff records, and an available head H. The turbine speed N will also be given, since it will usually be required to drive a generator at a prescribed rate. The choice of the type of turbine will then follow naturally after a calculation of specific speed, since we can observe in Table 3.2 that the ranges of specific speeds corresponding to peak efficiency are quite different for Pelton, Francis, and Kaplan Turbines. Thus the Pelton, or tangential-

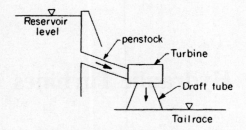

Figure 10.1 Hydraulic power plant arrangement.

flow, turbine is most efficient for $N_s < 0.3$, the Francis, or radial-flow, turbine is best for N_s between 0.3 and 2.0, and the Kaplan, or axial-flow, turbine is best for $N_s > 2.0$.

10.2 Pelton Wheel

The Pelton-type turbine is illustrated in Figure 1.3. For this turbine the penstock ends in a nozzle which creates a high-speed water jet. The latter impinges on vanes in the form of hemispheres or half-ellipsoids. The force on the vanes, created by deflection of the water through just less than 180°, drives the wheel around against the resistance of a load.

As indicated in Figure 10.2, the absolute velocity V_1 of the jet is the arithmetic sum of its relative velocity W_1 and the vane speed U. Leaving the vane at some small angle β_2, the tangential component of the absolute velocity is given by

$$V_{u2} = W_2 \cos \beta_2 - U \tag{10.1}$$

Thus the energy transfer can be expressed as

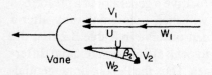

Figure 10.2 Vane velocity diagram for a Pelton wheel.

$$E = U(V_1 + W_2 \cos \beta_2 - U) \tag{10.2}$$

Since ideally

$$W_2 = W_1 = V_1 - U \tag{10.3}$$

we have

$$E = U(1 + \cos \beta_2)(V_1 - U) \tag{10.4}$$

If a velocity coefficient K is defined to account for the retardation of W_2, then (10.4) becomes

$$E = U(1 + K \cos \beta_2)(V_1 - U) \tag{10.5}$$

The hydraulic efficiency is the ratio of energy transfer to available energy gH. The latter is almost entirely converted to jet kinetic energy, i.e., $gHC = V_1^2/2$, so that

$$\eta_H = \frac{2EC}{V_1^2} \tag{10.6}$$

where C is a head loss coefficient.

Differentiation with respect to U/V_1 leads to the result that maximum hydraulic efficiency corresponds to

$$\frac{U}{V_{1opt}} = \frac{1}{2} \tag{10.7}$$

The shaft power P of the Pelton turbine may be estimated from the foregoing relations if leakage and mechanical losses are also considered. The latter losses are usually small, e.g., 3 to 5 percent, and are accounted for in the volumetric and mechanical efficiencies η_v and η_m, as with pumps. Thus we can write

$$P = \eta_m \eta_v \dot{m} E \tag{10.8}$$

or, in terms of volume flow rate and head,

$$P = \eta_m \eta_v \eta_H \rho Q g H \tag{10.9}$$

in which the product of three efficiencies is usually denoted by η, the overall efficiency. It should be noted that the flow rate also depends on the head, i.e.,

$$Q = \left(\frac{\pi}{4}\right) d^2 C (2gH)^{1/2} \tag{10.10}$$

where C is a coefficient which accounts for head loss due to friction in the nozzle, control valve, and penstock, and d is the jet diameter. Of course, several jets may be used around the periphery of the wheel, in which case the flows would be additive.

The ratio of wheel diameter D to jet diameter d varies from 6 to 25. The buckets are positioned close together to avoid spillage, the number of buckets varying from 20 to 30 per wheel. Such designs result in overall efficiencies of 80 to 90 percent.

10.3   Francis Turbine

The Francis turbine is a radial-flow reaction machine, much like a centrifugal pump with the flow direction reversed. Such a turbine is shown schematically in Figure 1.4. Flow from the penstock enters a spiral casing, which distributes water to the adjustable guide vanes located around the turbine wheel. The water leaves the guide vanes at angle α_1 and speed V_1, as depicted in Figure 10.3. Ideally, the vanes are designed so that the exit velocity V_2 has no whirl component, i.e., it is axial. Thus energy transfer is calculated from

$$E = U_1 V_1 \cos \alpha_1 \tag{10.11}$$

The shaft power P will again be calculated from (10.9) with hydraulic efficiency defined by

$$\eta_H = \frac{U_1 V_1 \cos \alpha_1}{gH} \tag{10.12}$$

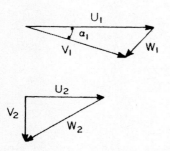

Figure 10.3 Velocity diagram for a Francis turbine.

and flow rate calculated from

$$Q = \frac{Cb\pi D}{(2gH)^{1/2}}$$

(10.13)

where b denotes axial width at the inlet and C denotes a velocity coefficient which depends on frictional resistance along the entire flow path from reservoir to tail race. Typically, the coefficient C is in the range 0.6 to 0.7.

A Cordier diagram, such as that appearing in Figure 3.3, can be used to determine a suitable wheel diameter D. This is obtained from the specific diameter after the Cordier diagram is entered for a known specific speed. The desired flow rate is assured by determining the appropriate axial width b_1 of the runner from (10.13).

Figure 10.4 illustrates differences in vane design arising from differences of the specific speeds of Francis runners (rotors). Low specific speeds imply low flow rates, which result in a smaller axial width b_1 and angle α_1, as well as a smaller exit diameter at the base of the rotor. Water leaves the high-specific-speed vane with a large axial component, in contrast to the large radial component present at the vane exit of a low-specific-speed runner. The inlet vane angle β_1 is also increased and can be as much as 135°. The lower edge of the high-N_s vane must be twisted to provide axial discharge velocities at varying blade speed. The rotor exit diameter is as large as, or larger than, the mean inlet diameter, as shown in Figure 10.4, to accommodate the higher flow rate associated with larger values of N_s.

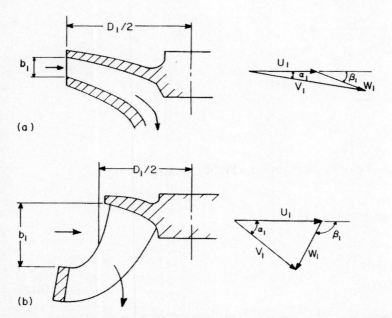

Figure 10.4 Francis turbine runner designs: (a) low specific speed; (b) high specific speed.

As indicated in Figure 1.4, the connection between the turbine casing and the tail race, called a draft tube, is installed to conserve the available head between these two levels. The draft tube accomplishes a diffusion as well, and the exit velocity from the system is thus reduced, improving overall efficiency.

10.4 Kaplan Turbine

The axial-flow hydraulic turbine is the third type of important hydraulic turbine. It is used primarily for low-head applications, i.e., for $H < 150$ ft. Such a turbine is shown in Figure 1.8. As is seen in this schematic representation, adjustable guide vanes are located around the inside of a volute casing, as with the Francis turbine. Also, the vanes of the propeller-type runner correspond to the lower part of the high-N_s Francis turbine vanes, where axial flow predominates. If the airfoil-shaped vanes are adjustable, the turbine is a Kaplan type; otherwise it is called a propeller type.

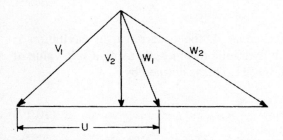

Figure 10.5 Velocity diagram for a Kaplan turbine.

The same principles previously applied to axial-flow steam and gas turbines apply to this machine, except that the fluid is treated as incompressible. Uniform axial velocity and the free-vortex variation of whirl velocity can be applied in the analysis of these blades as well. Whirl is imparted to the water in the inlet guide vanes, and it is deflected through a small angle by the blade. The exit velocity has no whirl component (see Figure 10.5) ideally, so that, as with the Francis turbine, we have

$$E = UV_{u1} \tag{10.14}$$

$$\eta_H = \frac{UV_{u1}}{gH} \tag{10.15}$$

The flow rate can be expressed in terms of a hub diameter D_h, a tip diameter D_t, and a coefficient C_k, which accounts for head loss due to friction. Thus we may write

$$Q = \frac{\pi}{4}(D_t^2 - D_h^2)C_k(2gH)^{1/2} \tag{10.16}$$

According to Kadambi and Prasad (1977), the coefficient C_k can vary from 0.35 to 0.75. Power to the turbine shaft is calculated from (10.9), as before.

10.5   Cavitation

As previously discussed in Chapter 4, the occurrence of cavitation is avoided in pumps, and also in hydraulic turbines, by using a safe value of the net positive suction head, NPSH. This is defined by

$$NPSH = H_{atm} + H_t - H_{vap} \tag{10.17}$$

where all terms in (10.17) are expressed in feet of liquid flowing, and the subscripts refer to atmospheric, total, and vapor pressures. H_t is the total (static plus dynamic) head, referred to the centerline of the turbine or pump. If the turbine has a draft tube, it is negative. Thus NPSH is a measure of the difference between the absolute static pressure at the turbine blade and the vapor pressure of the water in the turbine. If the NPSH falls below a certain critical value, the lowest static pressure in the turbine can be equal to the vapor pressure of the water, and cavitation can likewise occur.

An important ratio, known as Thoma's cavitation parameter σ, is defined by

$$\sigma = \frac{NPSH}{H} \tag{10.18}$$

Critical values of σ, corresponding to incipient cavitation, have been correlated with specific speed N_s defined as $NP^{1/2}/H^{5/4}$. For the Francis turbine Shepherd (1956) gives the critical Thoma parameter correlation as

$$\sigma = 0.625 \left(\frac{N_s}{100} \right)^2 \tag{10.19}$$

in which specific speed is calculated using rpm, hp, and ft units for N, P, and H, respectively. Similarly, the critical value of σ for Kaplan turbines is given as

$$\sigma = 0.28 + \frac{1}{7.5} \left(\frac{N_s}{100} \right)^3 \tag{10.20}$$

Values of σ calculated from flow conditions should always exceed values of σ calculated from (10.19) and (10.20) in order to assure avoidance of cavitation.

References

Daugherty, R. L. 1920. *Hydraulic Turbines*. McGraw-Hill, New York.
Esser, C. and J. H. T. Sun. 1989. *Hydraulic Turbines. Standard Handbook of Powerplant Engineering*. McGraw Hill, New York.
Kadambi, V. and M. Prasad. 1977. *An Introduction to Energy Conversion. Vol. 111, Turbomachinery*. Wiley, New York.
Shepherd, D. G. 1956. *Principles of Turbomachinery*. Macmillan, New York.
Wagner, H. T., Fischer, K. J., and J. D. v. Frommann. 1981. *Strömungs- und Kolbenmaschinen*. Viewag, Braunschweig.
Wislicenus, G. F. 1965. *Fluid Mechanics of Turbomachinary*. Dover, New York.

Problems

10.1 Develop an expression for design torque (maximum efficiency) for the Pelton turbine in terms of wheel diameter and jet characteristics.

10.2 For the turbine of Problem 10.1, H = 100 ft of water, β_2 =0 °, K = 1, jet diameter = 6 in., C = 0.94, η = 0.9, and N = 120 rpm. Find the following for maximum efficiency:

    a) flow rate
    b) power
    c) torque
    d) wheel diameter

10.3 A hydraulic turbine is to be designed to produce 36,600 hp at 93.7 rpm under a 54 ft head. A model of the turbine is to be designed to produce 50 hp at 554 rpm. If model efficiency is assumed to be 88 percent, find

    a) diameter of the model
    b) diameter of the turbine (prototype)
    c) type of turbine

11 Wind Turbines

11.1 Introduction

The natural motion of atmospheric air provides us with an excellent source of available energy for conversion to useful work. The moving rotor of a wind turbine arranged with a horizontal or vertical axis provides the means of energy conversion.

Wind turbines have been utilized by humans for hundreds of years for pumping water, grinding grain, and even generating electricity, but a renewed interest in the ancient device had been observed in recent years. The variability of the winds and the availability of cheap electrical energy has delayed the development of efficient, scientifically designed machines.

Csanady (1964) classifies the wind turbine as an extended turbomachine; the absence of a casing results in the rotor's effect extending to distant points in the fluid. It requires no nozzles to accelerate fluid, since the fluid is already moving. Although it lacks these conventional components, it can be analyzed as a turbomachine by applying momentum and energy principles to fluid passing through the rotor.

If fluid passes through the rotor in a direction parallel to the axis of the rotor, it can be called a horizontal-axis wind turbine. If the direction is normal to the axis, it is a vertical-axis machine. The horizontal-axis turbine may be compared with the axial-flow, or propeller, type of hydraulic turbine, but the vertical-axis type differs from the radial-flow turbine in that the flow is across the rotor rather than radially inward. Figures 11.1 and 11.2 show typical arrangements for these two types of wind turbines.

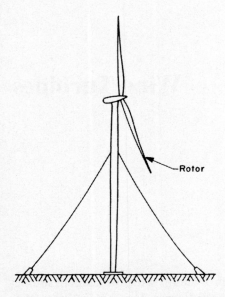

Figure 11.1 Horizontal-axis wind turbine.

In what follows an attempt will be made to discuss briefly the aero-dynamic design of the basic horizontal and vertical types of wind turbines. There are many considerations which are not discussed in this text, such as speed control, energy storage, and structural design. Generally, the wind machine must be designed to operate continuously at constant speed with various wind and weather conditions and to drive a generator or pump to supply immediate and future energy needs. Schemes for meeting these and other design requirements are discussed in books by De Renzo (1979), Bossel (1977), Calvert (1979), and Cheremisinoff (1978). Larger wind energy machines are discussed by Logan and Ditsworth (1986).

11.2   Actuator Theory

If one imagines the turbine rotor to be replaced by a disk of area A which can extract energy from the stream, as shown in Figure 11.3, then a useful analytical result follows through application of the basic equations. The theory has been developed by Glauert (1959) for propellers and by Betz

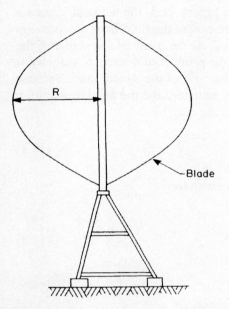

Figure 11.2 Vertical-axis wind turbine.

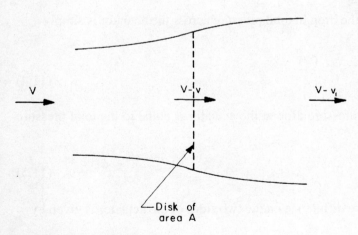

Figure 11.3 Actuator disk.

(1966) for windmills. As indicated in Figure 11.3, the actuator causes a divergence of streamlines and a deceleration of the fluid from an upstream speed V to a downstream speed $V - v_1$. At the plane of the actuator, the speed is $V - v$. The mass flow rate is the product of density ρ, velocity at the actuator $V - v$, and cross-sectional area A of the stream tube. Because ambient pressure prevails far from the actuator, and the axial momentum difference is simply v_1, we can express the drag by

$$D = \rho(V - v)Av_1 \tag{11.1}$$

Since the total pressure upstream p_0 is given by

$$p_0 = p_\infty + \frac{\rho V^2}{2} \tag{11.2}$$

and that downstream by

$$p'_0 = p_\infty + \frac{\rho(V - v_1)^2}{2} \tag{11.3}$$

we can see that the drop in total pressure across the actuator is simply

$$p_0 - p'_0 = \rho V v_1 - \frac{\rho v_1^2}{2} \tag{11.4}$$

Since the static pressure drop at the actuator is equal to the total pressure drop, we have

$$p_0 - p'_0 = p - p' \tag{11.5}$$

Thus the pressure difference on the two sides of the actuator is given by

$$p - p' = \rho v_1 \frac{V - v_1}{2} \tag{11.6}$$

The drag force on the actuator can then be expressed as

$$D = (p - p')A \tag{11.7}$$

or

$$D = \rho A v_1 \frac{V - v_1}{2} \tag{11.8}$$

Equating (1 1 . 1) and (11.8) yields

$$v_1 = 2v \tag{11.9}$$

Substituting (11.9) into (11.8) results in

$$D = 2\rho A v(V - v) \tag{11.10}$$

If the efficiency η of the wind turbine is defined by

$$\eta = \frac{T\Omega}{DV} \tag{11.11}$$

where T is the torque and Ω is rotational speed, then the efficiency compares the actual shaft power produced to the "thrust power" available from the stream. The actual retardation of the stream is less than DV, and it is derived as the rate of decrease of kinetic energy:

$$\Delta \dot{K} E = \rho(V - v)A \left(\frac{V^2}{2} - \frac{(V - v_1)^2}{2} \right) \tag{11.12}$$

In simplest form (11.12) reduces to $D(V - v)$, which is somewhat less than DV. Since this energy ideally is transferred to the rotor, we have

$$T\Omega = D(V - v) \tag{11.13}$$

The efficiency expression can then be rendered as

$$\eta = 1 - \frac{v}{V} \tag{11.14}$$

To determine the efficiency corresponding to maximum power, (11.14) is used to eliminate v in (11.10). The torque-speed product is replaced by power P; thus

$$P = T\Omega \tag{11.15}$$

and (11.11) becomes

$$P = 2\rho A V^3 \eta^2 (1 - \eta) \tag{11.16}$$

Differentiation of (11.16) yields

$$\frac{dP}{d\eta} = 2\rho A V^3 [2\eta(1 - \eta) - \eta^2] \tag{11.17}$$

Setting the derivative equal to zero gives one the optimum value of η, namely,

$$\eta_{opt} = \frac{2}{3}$$

or

$$\frac{v}{V} = \frac{1}{3} \tag{11.18}$$

Putting (11.18) into (11.16) gives the maximum wind turbine power as

$$P_{max} = \frac{8}{27} \rho A V^3 \tag{11.19}$$

Equation (11.19) is a well-known result and is frequently used to estimate the upper limit of power attainable from a wind turbine of area A in a wind of speed V and density ρ.

Actuator theory provides a useful model for the design of horizontal-axis machines. It must be extended to include the tangential component of velocity associated with energy transfer. This is accomplished in the next section, where the theory is related to the aerodynamic design of a rotor.

LOGAN   **235**

11.3 Horizontal-Axis Machines

Since the rotor imparts angular momentum to the air as it passes between
the blades, a change in whirl velocity ΔV_u must be indicated by the theory.
The change ΔV_u is expected to be proportional to the blade speed U, which
is the product of rotational speed Ω and radial position r. This may be
expressed as

$$\Delta V_u = 2a'\Omega r \tag{11.20}$$

where the coefficient a' accounts for the difference between the speeds of
blade and fluid.

The coefficient a' is used by Wilson and Lissaman (1974), who also
define the coefficient a by the expression for wind speed u at the actuator as

$$u = (1 - a)V \tag{11.21}$$

The actuator theory has shown that $a = \frac{1}{3}$ for maximum power, since a is
identical with v/V in (11.18). Wilson and Lissaman (1974) show that the
turbine delivers power for $0 < a < 0.5$. The theoretical value is helpful in
that $a = \frac{1}{3}$ can be chosen as a first approximation in design calculations.

Referring to Figure 11.4, we see that the velocity triangle is easily
constructed from a knowledge of radial position r, blade rotational speed
Ω, wind speed V, and the coefficient a. Normally, the designer would be
given values for wind speed V and rotor speed Ω. The solution of Glauert
for the optimum actuator disk given by Wilson and Lissaman (1974) is
useful in obtaining approximate values of the parameter a at each radial
position along the rotor blade. Glauert's solution is

$$\frac{r\Omega}{V} = \left(\frac{1-a}{1-3a}\right)^{1/2}(4a - 1) \tag{11.22}$$

from which the dimensionless group $r\Omega/V$ can be determined for any value
of the parameter a. The relation can be employed to make tables or graphs
for practical use. Figure 11.5, based on (11.22), is such a plot. It is clear
from Figure 11.5 that $a \approx \frac{1}{3}$ over most of a practical blade. The axial
component of the absolute velocity at the rotor is then approximately
two-thirds of the wind speed.

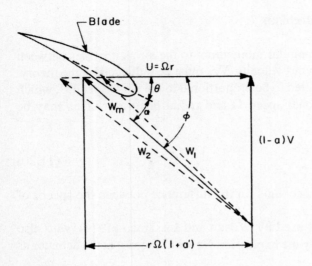

Figure 11.4 Velocity diagram for a horizontal-axis turbine.

The Glauert theory also relates the parameter a', defined in (11.20), to the parameter a by the relation

$$a' = \frac{1 - 3a}{4a - 1} \tag{11.23}$$

Calculation of a' completes the required operations to ensure the construction of a velocity diagram such as that shown in Figure 11.4 at any radial position r along the blade length.

To determine the blade pitch angle θ at any radial position r, we first calculate the angle ϕ from the relation

$$\tan \phi = \frac{1 - a}{(1 + a')(r\Omega/V)} \tag{11.24}$$

Next, we must consult a reference such as Abbott and von Doenhoff (1959) for the determination of a suitable angle of attack α. Airfoil lift curves (a plot of c_L versus α) indicate the stalling value of α. The angle of attack for design purposes will be less than this value, but near it, to get the largest possible force from the wind. Selection of this value of α for design also

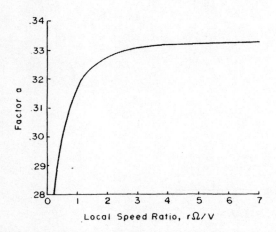

Figure 11.5 Dependence of parameter a on the local speed ratio.

provides us with lift and drag coefficients at that section. Calculation of the blade pitch angle can be made easily as well.

Bossel (1977) shows data which indicate that the power coefficient, defined by

$$c_p = \frac{2P}{\rho V^3 A}$$

(11.25)

depends upon blade number and the tip-speed ratio. An attempt to make a rough correlation of the tip-speed ratio corresponding to maximum c_p as a function of blade number has been made, and the results are shown in Figure 11.6. The 24-blade American windmill, having a solidity of very nearly unity and a c_p of about 0.15, requires a tip-speed ratio of less than unity. A two-bladed, high-speed wind turbine, having a c_p slightly less than 0.5, on the other hand, has a tip-speed ratio of 6 or more. The need for simplicity, economy, and efficiency would indicate a choice of the higher tip speeds with correspondingly fewer blades.

A preliminary design of a rotor blade could be effected using the following approximation for the blade chord c

$$c = \frac{4\pi r \sin^2 \phi}{n_b c_L \cos \phi}$$

(11.26)

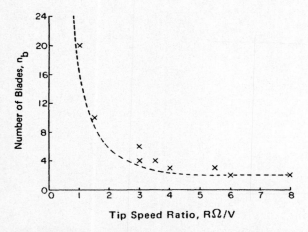

Figure 11.6 Variation of blade number with tip speed.

which follows from the Glauert theory presented by Wilson and Lissaman (1974). Using (11.26), a value of the chord of the airfoil section for the turbine blade can be determined at each radial position in terms of the airfoil lift coefficient and the velocity diagram angle ϕ applicable at that position r. The number of blades n_b would have been previously determined using the guidance of correlations such as those presented in Figure 11.6.

Lift and drag coefficients are utilized to compute torque for the design obtained. The tangential component of the blade force per unit length is given by

$$F_t = \frac{\rho W_m^2 c(c_L \sin \phi - c_D \cos \phi)}{2} \tag{11.27}$$

Torque on the rotor is computed by integrating F_t from the hub radius to the tip radius. Thus

$$T = n_b \int_{r_h}^{r_t} F_t r \, dr \tag{11.28}$$

which can be evaluated numerically. The turbine power follows from

$$P = T\Omega \tag{11.29}$$

As pointed out by Wilson and Lissaman (1974), the values of a, a', ϕ, α, c_L, and c_D must be refined before the final calculation of blade force are made. Since torque and thrust can be calculated from momentum or blade-element theory, internal consistency requires that

$$\frac{a'}{1+a'} = \frac{n_b c(c_L \sin\phi - c_D \cos\phi)}{4\pi r \sin 2\phi} \tag{11.30}$$

and

$$\frac{a}{1-a} = \frac{n_b c(c_L \cos\phi + c_D \sin\phi)}{8\pi r \sin^2\phi} \tag{11.31}$$

be satisfied. An iterative procedure can be carried out using the initially chosen values of a and a' to start the calculation of ϕ, α, c_L, c_D, and new values of a and a' from (11.30) and (11.31). After convergence is obtained the set of values may be used to calculate blade force and torque.

The method has been improved by Tangler (1987) to include the effects of wind crossflow and blade stall.

11.4 Vertical-Axis Machines

As noted by Blackwell (1974), the vertical-axis turbine of the type shown in Figure 11.2 operates equally well at any wind orientation. The airfoil-shaped blades follow a circular path, as indicated in Figure 11.7, and this motion results in a continually changing angle of attack α.

It is necessary that the force on the blade have a component in the direction of its motion to sustain the rotation of the rotor. The velocity triangles for a blade in the four quadrants are shown in Figure 11.7. A lift vector is sketched on each diagram perpendicular to the relative velocity vector. The diagrams clearly indicate that there exists a tangential component of the lift vector L in all four quadrants. Thus a favorable torque is produced regardless of blade position. Wilson and Lissaman (1974) show that the torque per blade can be estimated by the expression

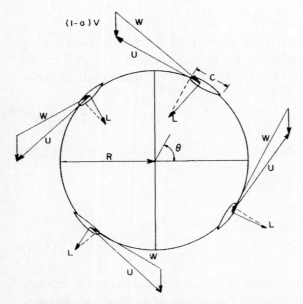

Figure 11.7 Velocity diagram for a vertical-axis turbine.

$$T = \rho\pi cRV^2(1 - a)^2 \sin^2\theta \tag{11.32}$$

where a is given by

$$a = \frac{n_b cR\Omega\,|\sin\theta|}{2RV} \tag{11.33}$$

and R is the maximum radius of the blade with respect to the axis of rotation, as shown in Figure 11.2. The airfoil chord is denoted by c, and the angular position θ is as shown in Figure 11.7. The power is obtained from (11.29).

Blackwell (1974) finds that maximum efficiency is obtained when the tip-speed ratio $R\Omega/V$ is around 6. The velocity diagrams of Figure 11.7 indicate a tip-speed ratio of about 4. Blackwell finds that vertical-axis turbines of this type fail to deliver power for $R\Omega/V < 3$.

Two other types of vertical-axis machines which have achieved fame are the Savonius (S-shaped blade) and the Madaras (cylinders). For a discussion of these and other types the reader is referred to a book by Simmons (1975).

References

Abbott, I. H., and A. E. von Doenhoff. 1959. *Theory of Wing Sections.* Dover, New York.

Betz, A. 1966. *Introduction to the Theory of Flow Machines.* Pergammon, Oxford.

Blackwell, B. F. 1974. *The Vertical-Axis Wind Turbine, "How It Works".* Sandia Albuquerque Laboratories Report No., SLA-74-0 160, Albuquerque, New Mexico.

Bossel, V. 1977. *Energie vom Wind.* Deutsche Gesellschaft für Sonnenenergie, Munich.

Calvert, N. G. 1979. *Windpower Principles: Their Application on the Small Scale.* Griffin, London.

Cheremisinoff, N. P. 1978. *Fundamentals of Wind Energy.* Ann Arbor Science, Ann Arbor, Michigan.

Csanady, G. T. 1964. *Theory of Turbomachines.* McGraw-Hill, New York.

De Renzo, D. J. 1979. *Wind Power.* Noyes Data Corporation, Park Ridge, California.

Glauert, H. 1959. *The Elements of Aerofoil and Airscrew Theory.* Cambridge Press, Cambridge, Massachusetts.

Logan, E. and R. L. Ditsworth. 1986. *Technical Assessment of Large Wind Energy Conversion Systems.* Navy Civil Engineering Laboratory Report, Port Hueneme, California.

Simmons, D. M. 1975. *Wind Power.* Noyes Data Corporation, Park Ridge, California.

Tangler, J. L. 1987. *A Horizontal Axis Wind Turbine Performance Prediction Code for Personal Computers.* Solar Energy Research Institute Report.

Wilson, R. E., and P. B. S. Lissaman. 1974. *Applied Aerodynamics of Wind Power Machines.* Oregon State University Report No. NSF-RA-N-74-113.

Problems

11.1 The Boeing MOD-2 wind turbine has a blade diameter of 300 ft and reaches a rated power of 2.5 MW at 27.5 mph. What is its power coefficient under these conditions?

11.2 For a horizontal-axis wind turbine operating at a = ⅓, show that the entropy increases through the actuator disk is given by

$$\Delta S = \frac{8}{9} R \frac{(\frac{1}{2} \rho V^2)}{P_\infty}$$

Sketch the flow process from upstream, through the actuator disk, and into the wake on a T-S diagram.

11.3 For a vertical-axis wind turbine, momentum conservation requires that the wake of the turbine be at some angle α to the free-stream velocity. Find α in terms of the other parameters for a vertical-axis machine.

Appendix A

Table 1 Useful Equivalents

Quantity	Original unit	Equivalent
Flow	1.0 cfs (cubic feet/sec)	448.8 gpm (gallons/min)
Specific Energy	1.0 ft^2/s^2	1.0 ft-lb$_F$/slug
Mass	1.0 slug	32.174 lb$_m$
Rotational Speed	1.0 rad/s	9.54929 rpm (revolutions/min)
Kinematic Viscosity	1.0 ft^2/s	92,903 cs (centistokes)
Specific Speed	1.0 (unitless)	$2732.9 \dfrac{\text{rpm(gpm)}^{½}}{\text{ft}^{¾}}$
Pressure	1.0 inch water	5.2 psf (lb$_f$/ft^2)

Table 2   Pump Efficiency as a Function of Specific Speed and Capacity

N_s Dimensional specific speed	Capacity Q (gpm)								
	30	50	100	200	300	500	1000	3000	10000
200	0.21	0.24	0.29	–	–	–	–	–	–
300	0.31	0.34	0.40	–	–	–	–	–	–
400	0.38	0.42	0.48	–	–	–	–	–	–
500	0.43	0.47	0.54	0.59	0.61	0.64	0.67	0.70	0.73
600	0.47	0.50	0.57	0.62	0.65	0.67	0.70	0.73	0.76
800	0.52	0.56	0.63	0.67	0.70	0.72	0.75	0.78	0.82
1000	0.55	0.60	0.66	0.70	0.73	0.76	0.78	0.82	0.85
1200	0.57	0.62	0.68	0.72	0.75	0.78	0.81	0.84	0.87
1400	0.59	0.63	0.70	0.73	0.76	0.79	0.82	0.85	0.87
1600	0.60	0.64	0.70	0.74	0.77	0.80	0.83	0.86	0.88
1800	0.60	0.65	0.71	0.74	0.77	0.80	0.83	0.86	0.88
2000	0.60	0.65	0.71	0.75	0.77	0.81	0.84	0.87	0.89
3000	0.60	0.65	0.71	0.75	0.77	0.81	0.84	0.87	0.89

Source: Constructed from graphs in the *Pump Handbook*. Karassik I. J. et al (Eds.) McGraw-Hill, New York, 1976.

Table 3 Compressor Specific Diameter as a Function of Specific Speed and Efficiency

Dimensional specific speed	Compressor efficiency				
	0.40	0.50	0.60	0.70	0.80
50	2.42	2.65	2.91	–	–
60	1.94	2.14	2.26	–	–
65	1.77	1.92	2.02	2.14	–
70	1.66	1.82	1.89	1.96	–
80	1.44	1.55	1.63	1.68	–
85	1.36	1.48	1.53	1.57	1.70
90	1.30	1.39	1.43	1.46	1.59
100	1.16	1.25	1.29	1.32	1.41
110	1.07	1.14	1.17	1.21	1.29
120	1.00	1.06	1.10	1.15	1.22
130	0.91	1.00	1.03	1.08	1.18
140	0.87	0.96	1.00	1.06	–
150	0.83	0.94	1.00	1.07	–
160	0.80	0.91	1.00	1.04	–
170	0.80	0.91	1.00	1.11	–
180	0.80	0.91	1.00	–	–
190	0.79	0.91	1.01	–	–
200	0.79	0.91	–	–	–

Source: Scheel, L. F. 1972. *Gas Machinery.* Gulf Publishing Co., Houston.

Appendix B

Derivation of Equation for Slip Coefficient

The relative eddy (Figure B-1) at the impeller tip is located between adjacent vanes, has diameter d and rotates at the impeller rotation rate N, but in the opposite direction to that of the impeller. The triangle ABC has one vertex at point A on the periphery of the impeller. The tangent AB to the vane forms the vane angle β_2 with the tangent AC to the circular wheel periphery at point A. The length of the leg BC is approximated by the eddy diameter d, and the hypotenuse AC is taken as equal to the arc length AD. The slip $V_{u2} - V_{u2}'$ or ΔW_u is calculated as the speed Nd/2 of a point on the

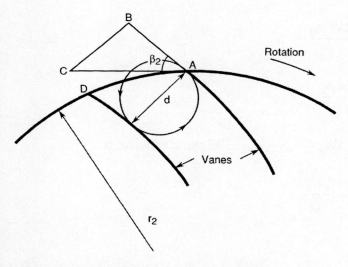

Figure B-1 Relative eddy at impeller tip.

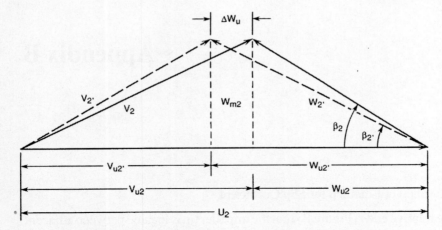

Figure B-2 The effect of slip on the exit velocity diagram.

circumference of the relative eddy, in which only solid body rotation is present. The foregoing assumptions lead to the following formulations:

$$\sin \beta_2 = \frac{BC}{AC} = \frac{dn_B}{2\pi r_2}$$

$$Nr_2 \sin \beta_2 = \frac{\Delta W_u n_B}{\pi}$$

$$\Delta W_u = \frac{\pi U_2 \sin \beta_2}{n_B}$$

From Figure B-2, $\Delta W_u = V_{u2} - V_{u2'}$. The slip coefficient is

$$\mu_s = \frac{V_{u2'}}{V_{u2}} = 1 - \frac{\Delta W_u}{V_{u2}}$$

and, finally,

$$\mu_s = 1 - \frac{\pi U_2 \sin \beta_2}{n_B V_{u2}}$$

Appendix C

Formulation of Equation for Hydraulic Loss in Centrifugal Pumps with Backward-Curved Vanes

According to Csanady (1964) the hydraulic loss in a centrifugal pump can be represented by the relation

$$gH_L = \frac{k_d V_{2'}^2}{2} + \frac{k_r W_{2'}^2}{2}$$

(C1)

where $V_{2'}$, and $W_{2'}$, denote the absolute and relative velocities, as are shown in the velocity diagram of Figure B-2, and k_r and k_d represent the loss coefficients for rotor and diffuser, respectively. Referring to Figure B-2 in Appendix B, it is seen that

$$\sin^2 \beta_{2'} = \frac{W_{m2}^2}{W_{2'}^2}$$

(C2)

and that

$$W_{2'}^2 = W_{m2}^2 + (U_2 - V_{u2'})^2$$

(C3)

Noting that the actual and ideal flow coefficients are equal, we can write

$$\varphi_2 = \varphi_{2'} = \frac{W_{m2}}{U_2}$$

(C4)

Substitution of (C3) and (C4) into (C2) gives

$$\sin^2 \beta_{2'} = \frac{\varphi_2^2}{\varphi_2^2 + (1 - V_{u2'}/U_2)^2} \tag{C5}$$

The hydraulic efficiency is defined as

$$\eta_H = \frac{H}{H_{in}} = \frac{H_{in} - H_L}{H_{in}} \tag{C6}$$

It is clear from (C6) that the maximum efficiency corresponds to the minimum H_L/H_{in}. To obtain this form we divide (C1) by gH_{in}, i.e., by $U_2 V_{u2'}$ and we obtain

$$\frac{H_L}{H_{in}} = \frac{k_d V_{2'}^2 + k_r W_{2'}^2}{2 U_2 V_{u2'}} \tag{C7}$$

From Figure B-2 we observe that

$$V_{2'}^2 = W_{m2}^2 + V_{u2'}^2 \tag{C8}$$

Substituting (C2) and (C8) into (C7) yields

$$\frac{H_L}{H_{in}} = \frac{k_d W_{m2}^2 + k_d V_{u2'}^2 + k_r W_{m2}^2/\sin^2 \beta_{2'}}{2 U_2 V_{u2'}} \tag{C9}$$

Factoring, substituting (C4) into (C9) and rearranging gives the final form,

$$\frac{2H_L}{k_d H_{in}} = \frac{\varphi_2^2}{V_{u2'}/U_2} \left(1 + \frac{k_r/k_d}{\sin^2 \beta_{2'}} \right) + \frac{V_{u2'}}{U_2} \tag{C10}$$

Using Csanady's value of $\frac{1}{3}$ for k_r/k_d allows evaluation of the parameter on the left hand side of (C10), which is to be minimized, for each pair of assumed values of φ_2 and $Vu_{2'}/U_2$. Determination of the optimum value of $V_{u2'}/U_2$ is a straightforward task and is carried out in Problem 4.13.

Optimum values of V_{u2}/U_2 are in the range between 0.5 and 0.55 for typical values of flow coefficient. There is a corresponding optimum value of β_2 for each flow coefficient; it varies from 5.7° for $\varphi_2 = 0.05$ to 23.5° at $\varphi_2 = 0.20$.

Reference

Csanady, G. 1964. *Theory of Turbomachines*. McGraw-Hill, New York.

Appendix D

Viscous Effects on Pump Performance

Table 1 Viscosities of Liquids at 70°F

Liquid	Kinematic viscosity (centistokes)
Ethylene glycol	17.8
Water	1.0
Gasoline	0.67
Kerosene	2.69
Engine oil	165
Glycerine	648

Table 2 Correction Factors for Oil with Kinematic Viscosity of 176 Centistokes

Water capacity (gpm)	Factor	Head (ft)		
		15	100	600
100	c_E	0.39	0.49	0.60
	c_Q	0.80	0.88	0.93
	c_H	0.82	0.87	0.90
1000	c_E	0.60	0.67	0.72
	c_Q	0.93	0.96	0.97
	c_H	0.90	0.92	0.93
5000	c_E	0.72	0.77	0.83
	c_Q	0.97	0.98	1.0
	c_H	0.93	0.95	0.96

Source: Karassik, I. J. et al. 1976. *Pump Handbook*. McGraw-Hill, New York.

Appendix E

Comparison of Formulas for Compressor Slip Coefficient, $\mu_s = V_{u2'}/V_{u2}$

Originator	Reference	Formula
Balje	1	$1 - \dfrac{0.75\pi \sin \beta_2}{n_B}$
Busemann	6	$1 - \dfrac{2.4}{n_B}$ ($\beta_2 = 90°$ only.)
Eck	8	$\left[1 + \dfrac{2 \sin \beta_2}{n_B[1 - (D_{1S}/D_2)]}\right]^{-1}$
Pfleiderer	7	$\left[1 + \dfrac{8(k + 0.6 \sin \beta_2)}{3n_B}\right]^{-1}$ ($0.55 < k < 0.68$)
Stodola	2, 4, 5	$1 - \dfrac{\pi}{n_B}\left[\dfrac{\sin \beta_2}{1 - \varphi_2 \cot \beta_2}\right]$
Stanitz	2, 3, 4	$1 - \dfrac{0.63\pi}{n_B}\left[\dfrac{1}{1 - \varphi_2 \cot \beta_2}\right]$

References

1. Balje, O. E. 1981. *Turbomachines.* John Wiley & Sons, New York.
2. Boyce, M. P. 1982. *Gas Turbine Engineering Handbook.* Gulf Publishing Co., Houston.
3. Cohen, H., et al. 1987. *Gas Turbine Theory.* Longman, Essex.
4. Dixon, S. L. 1975. Fluid Mechanics. *Thermodynamics of Turbomachinery.* Pergamon, Oxford.
5. Ferguson, T. B. 1963. *The Centrifugal Compressor Stage.* Butterworths, London.
6. Harman, R. T. C. 1981. *Gas Turbine Engineering.* John Wiley & Sons, New York.
7. Pfleiderer, C. 1949. *Die Kreiselpumpen.* Springer-Verlag, Berlin.
8. Wilson, D. G. 1984. *The Design of High-Efficiency Turbomachinery and Gas Turbines.* MIT Press, Cambridge.

Appendix F

Table F-1 Molecular Weight of Selected Gases

Gas	Molecular weight
Air	28.966
Oxygen	32.00
Nitrogen	28.016
Carbon monoxide	28.010
Carbon dioxide	44.010
Hydrogen	2.016
Methane	16.042
Ethane	30.068
Propane	44.094
n-Butane	58.120

Source: Scheel, L. F. 1972. *Gas Machinery.* Gulf
Publishing Co., Houston.

Index